The Creative Conscience
as Human Destiny

american
university
studies

Series V
Philosophy

Vol. 196

PETER LANG
New York • Washington, D.C./Baltimore • Bern
Frankfurt am Main • Berlin • Brussels • Vienna • Oxford

Edward H. Strauch

The Creative Conscience as Human Destiny

PETER LANG
New York • Washington, D.C./Baltimore • Bern
Frankfurt am Main • Berlin • Brussels • Vienna • Oxford

Library of Congress Cataloging-in-Publication Data

Strauch, Edward H.
  The creative conscience as human destiny / Edward H. Strauch.
    p. cm. — (American university studies. V, Philosophy; v. 196)
    Includes bibliographical references and index.
  1. Life. 2. Evolution. I. Title. II. Series: American university studies.
    Series V, Philosophy; v. 196.
      BD431 .S855   128—dc22   2003022532
          ISBN 0-8204-6832-0
          ISSN 0739-6392

Bibliographic information published by **Die Deutsche Bibliothek**.
Die Deutsche Bibliothek lists this publication in the "Deutsche
Nationalbibliografie"; detailed bibliographic data is available
on the Internet at http://dnb.ddb.de/.

The paper in this book meets the guidelines for permanence and durability
of the Committee on Production Guidelines for Book Longevity
of the Council of Library Resources.

© 2004 Peter Lang Publishing, Inc., New York
275 Seventh Avenue, 28th Floor, New York, NY 10001
www.peterlangusa.com

All rights reserved.
Reprint or reproduction, even partially, in all forms such as microfilm,
xerography, microfiche, microcard, and offset strictly prohibited.

Printed in Germany

*To those seeking a life philosophy
based on the wisdom of nature.*

CONTENTS

ACKNOWLEDGMENTS xiii

INTRODUCTION xv

PART I
THE EVOLUTION OF LIFE

1 DOES BIOGENETIC INTELLIGENCE ACCOUNT FOR EVOLUTION? 3

PART II
A 2,500 YEAR HISTORY OF BEING AND BECOMING

2 BEING AND BECOMING 19
    The Principle of Plenitude 19
    The Principle of Continuity 21
    The Principle of Gradation 22
    The Meaning of Nature's Principles 23
    Toward a Comprehension of Nature and Evolution 24
    The Value of the History of Ideas 26
    Metaphor as Metaphysical Meaning 28
    Personification 29
    Symbol 29
    Friedrich Schiller 30
    A Biogenetic Basis for Philosophy 32

PART III
LIFE AS PURPOSE

3 NATURE'S POLARITIES AND WHAT THEY TEACH US 37
    Divergence and Convergence in Nature 40
    Divergence and Convergence in Human Nature 40
    Growth of Complexity Versus Simplification 42
    Dualism Versus Polarity 44
    Evolution as Polar Process 45
    Two Visions of Human Nature 47

4 NATURE'S PROCESSES SERVE A MUTUAL PURPOSE 49
    Morphogenesis and Symbiosis 53
    A Morphological View of Nature 57
    What Morphogenetic Nature Shows Us 59
        Plant Life 59
    Other Forms of Life: Insects, Fish, Animals 60
    Symbiosis in External Nature: Insects, Fish, Animals 62
    Other Manifestations of Symbiosis 63
    Microscopic Nature 65
    The Gene Connection 66
    Tentative Conclusions 70
        Morphogenesis as Polarity 70
        Symbiosis as Polarity 70
        Morphogenesis and Symbiosis as Dialectic 71
        The Mutual Purpose of Nature's Polarities 72

PART IV
INTELLIGENCE IN NATURE AND HUMANITY

5 INTELLIGENCE IN NATURE 77
    Thinking Learned from Nature 78
    Mothers in Nature 80
    Forms of Survival Intelligence 81
    Survival Intelligence in Animal and Man 83
    Morphogenetic/Symbiotic Intelligence 84
    Morphogenetic and Symbiotic Modes of Reasoning 85

Morphogenetic Experience, Symbiotic Verification 87
Nature as Source of Human Imagination 87
The Evolutionary Increase in Human Intelligence 89
Simplification and Complexification in Nature 89
Our Mental Processes 90

6 Forms of Cultural, Rational and Natural Intelligence 93

Cultural Intelligence 93
Rational Intelligence 97
Induction and Deduction 97
Problem-Solving Intelligence 98
Forms of Biological Intelligence 99
    Homologous Intelligence 99
    Holistic Intelligence 100
    Homeostatic Intelligence 101
    Morphogenetic/Symbiotic Intelligence 101
    Ontogenetic Intelligence 103
    Evolutionary Intelligence 103
Philosophical Intelligence 103
A Tentative Summary 104

7 How Nature's Processes Created Human Intelligence 107

Introduction 107
Universal Processes in Nature 111
    Embryonic Development and Lifetime Transformation 111
    Human Evolution 112
Polarities Actualize the Evolution of the Body-Mind 114
    Divergence 114
    Convergence 115
Growth of Complexity Versus Simplification 116
    Toward Complexity 116
    Simplification 117
    Morphogenesis 118
    Symbiosis 121
Tentative Conclusions 122
    Evolution 122
    Mutualism 123

Divergence 123
Convergence 123
Morphogenesis 124
Symbiosis 125
The Significance of Nature's Processes
to the Human Mind 126

PART V
HUMAN EVOLUTION IN CULTURE, CREATIVITY AND CONSCIENCE

8 THE POLARITY OF HUMAN NATURE IN CULTURE 129
Introduction 129
Human Divergence and Convergence 130
    Divergence 131
    Convergence 132
Tentative Conclusions 137

9 THE BEGINNINGS OF HUMAN INGENUITY AND CONSCIENCE 139
Human Resourcefulness as the Basis of Survival 139
Laws, Rights, and Civilization 147
Conscience in Past Life Philosophies 150
The Enlightenment 159
Deism 160
Conclusion 160

PART VI
THE INTELLIGENT LIFE

10 A THIRD MILLENNIUM LIFE PHILOSOPHY 167

11 THE CREATIVE CONSCIENCE AS HUMAN DESTINY 179
Introduction:
Conflicting Interpretations of Evolution in History 179
The New Reality 181
The Question of Extinction 184
The Argument Thus Far 187
The Dynamic Polarity of Human Nature: Human Intelligence ... 187

The Morphogenetic/Symbiotic Conscience 188
The Morphogenetic Phase of the Creative Conscience 191
Human Conscience and Creativity 192
The Archetypal Destinies of Man and Woman 194
Nature's Influence on Human Character and Personality 195
Paul Diel's Interpretation of Psyche 197
The Subconscious 198
The Supraconscience 198
Morphogenesis and Symbiosis
in the Context of Diel's Psychology 198
The Consequence of Co-Evolution 199
The Supraconscience as Creative Conscience 199
Metaphor and Irony as Keys to Human Destiny 200

12 THE PLAN AND PURPOSE OF NATURE 205
Introduction 205
Hierarchy in Nature 206
Survival and the Acceleration of Intelligence 207
Intelligence as Consequence of Co-Evolution 208
Morphogenetic Human Nature 209
The Need for Creativity and Conscience 210
Nature's Processes Account for Our Mental Evolution 210
How Nature Teaches Us to Survive 211
Four Symbiotic Principles of Survival 212
The Symbiotic Conscience 212
Consequences of Human Evolution 213
On Remembering our Own Heritage....................... 215
On Moral Education 215
The Creative Conscience as Life Philosophy 216
Stages of Life and Culture............................... 218
The Characteristics of the Creative Conscience 219
Morphogenesis: Its Traits and Functions 219
Symbiosis: Its Traits and Functions 220

AFTERWORD: THE MUTUAL PURPOSE OF
  NATURE'S PROCESSES AND SOME FINAL CONCLUSIONS 221

NOTES 225

GLOSSARY 229
BIBLIOGRAPHY 233
TOPICAL ALPHA INDEX 237

ACKNOWLEDGMENTS

A noteworthy scholar of history, Dr. Frank P. King is a true friend, who tirelessly reviewed the manuscript, gave sound and germane advice, and encouraged my bent for philosophy.

Dr. Charles F. Herberger, Professor of English, whose comprehensive knowledge of Charles Darwin's theory of evolution helped me to use the theory judiciously, and who over the decades patiently guided my own modest, scholarly efforts.

Lorenz A. Mundstock, retired Professor of Philosophy, friend for more than fifty years, whose insights into social inequities and the absence of conscience of vested interests taught me valuable lessons about humanity's rights.

Ofelia P. Villagomez for her unwavering moral support, patience, dependability, and example of true devotion to one's religion.

To my family, long gone, for their love, trust, and faith in the future. They taught me self-discipline, dedication to ideals, concern for humankind, and the need to find fulfillment in meaningful, lifetime work.

Susan Peabody, of Camera Ready Copy, in Berkeley, California, whose computer skills, commitment to the task, and exceptional competence have made possible the preparation of this manuscript for publication.

The Peter Lang editorial staff especially have my gratitude. Justin Pelegano, Editorial Assistant, adeptly guided me through the first steps of the overall publication procedure. Phyllis Korper, Acquisitions Editor, gave the introductory chapters a judicious first reading, highlighted items needing correction, and provided invaluable, constructive criticism. Sophie Appel, Production Coordinator, equally deserves sincere thanks for her thoughtful and insightful guidance through all stages of the book production. I am grateful to everyone for their professional skill and experience. Otherwise, whatever failings the manuscript may have are entirely my own.

INTRODUCTION

Twentieth century ecology and microbiology seem to propose a new interpretation of evolution. They have discovered that Nature is animated by creative powers (*morphogenesis*) and powers of integration (*symbiosis*, which together appear to be the actual source of evolution. Evidently, Nature's Intelligence gave birth to human intelligence. Out of Nature's infinite Creative Conscience came the finite creative conscience of humanity.

If this hypothesis is to be proven, it must answer two fundamental questions. Does biogenetic intelligence account for human evolution? Do morphogenesis and symbiosis together actualize our evolution?

Microbiology has shown how change is energized and designed from within all forms of life. Living cells manifest both the power of generation and of consolidation. Cellular processes are responsible for every creature's genesis, and these microactivities seem to incarnate intelligence and purpose.

Similarly genes control and guide inheritable tendencies in all plants and animals. Moreover, gene complexes work in mutually cooperative ways, thus apparently acting with intelligence and purpose. Moreover, stages of individual genesis and of species evolution represent stages of evolving intelligence.

Throughout the living environment, evidence of intelligence is substantiated by the regularity, rhythm, patterns and cycles of Nature and in all life that manifests perfected form. By working synergistically with morphogenesis, symbiosis enables a life form to build on its successful experiments with life. Thus life forms evolve "conscience." Morphogenesis is consolidated into a symbiotic conscience.

The bio-logic of Nature transcends the rationale of past systems of reasoning. Nature's intrinsic creativity and self-organization require us to reevaluate our traditional ways of perceiving and interpreting reality.

Nature's morphogenetic intelligence is essentially explorative and creative. On the other hand, its symbiotic intelligence coalesces, coordinates and unites the experiences of an evolved life-form into a nature-born conscience. Hence the correspondence between Nature and human nature is unmistakable. This interaction of morphogenesis and symbiosis explains the emergence of human intelligence and the creative conscience. Both are endowed to humanity by Nature. In sum, it appears that biogenetic intelligence accounts for human evolution.

Over the past 2,500 years, philosophers have exposed various interpretations of the living universe. In antiquity and the Middle Ages, many thinkers considered plenitude to describe the self-sufficient perfection of the universe and the rationality of existence. Others insisted that continuity was a better explanation to describe the increasing perfection of Nature. Finally, the later thinkers maintained that God ensured the degrees of perfection and excellence in Nature in the principle of gradation. Thus both being and becoming explained the universe to be one vast chain of being.

In the nineteenth century, Charles Darwin called attention to two polar processes in nature: divergence and convergence. Divergence from an archetype took place with the evolution of descendants into families, genera and species. Convergence took place when species, without a common ancestor, assumed similar shapes in a common medium like the sea.

An analogous polarity exists in human nature via mental divergence and convergence. Divergence occurs when individuals create new modes of self-expression and self-realization. Convergence takes place when hominid groups face similar survival problems and resolve them. In essence, polar extremes seem to have served a definite purpose in fostering human evolution.

The reciprocity between nature's powers of generation and of integration into life-forms proves that nature's processes serve a mutual purpose to create and sustain viable forms of life. Indeed, microbiology and ecology provide evidence of the effectiveness of morphogenetic and symbiotic interactivity.

Nature also abounds in examples of the survival of intelligent insects, birds, animals, and humankind. To be resourceful and to learn is to survive.

In fact, survival intelligence was the first form of intelligence to be developed in nature. In more evolved forms, morphogenetic/symbiotic intelligence emerged. Humanoid creativity and integration of experience into meaning fostered human intelligence. Indeed, religion, culture and civilization came into being by these reciprocal processes which together effected the evolution of the human mind.

Out of our biological aptitudes, we gradually evolved homologous intelligence to investigate the shapes, forms and phenomena in nature. On the other hand, our holistic intelligence studied how details and parts revealed nature's wholeness and integrity. Similarly, our homeostatic intelligence sought to balance, harmonize and integrate our needs, desires, aptitudes and experiences. The human mind gradually learned to heed its own checks and balances.

All these forms of cultural, rational and biological intelligence became integrated in the creative conscience of humanity. Our biological heritage demonstrates that our cells, genes, life-support systems and brain incarnate millions of years of evolutionary experience. Our bioheritage brings with it a sense of the purposiveness inherent in all life-forms. All evolved designs and forms in nature serve a purpose of their own. At our evolved level, our purpose is to attain true conscience by realizing the essential purpose of our life.

To be sure, the meaning of conscience extends beyond any civilized notion of sin or guilt. Human conscience implies mutual aid among humankind, the readiness to help others survive.

Conscience evolves when new experience evaluates the old. Conscience also evolves when it simplifies the complex into clearer comprehension. Human conscience evolves when it learns to shape life's complexities into definite moral judgements.

The impulse to *divergence* that we inherited from nature impels us to pursue a variety of experiences whereas *convergence* guides us to concentrate our life energies to pursue worthwhile goals. Concentration teaches conscience the benefits of self-control, temperance and self-discipline. Conscience also urges us to harmonize intelligence, talents, and the need to survive by seeking to achieve life-fulfilling objectives. In combination with conscience, our in-born creativity created new laws, forms of government, order, morality and all actions of man's humanity to man. In sum, our mental processes replicate our natural evolution.

Nature's divergence and convergence probably influenced the history of human culture. This is manifest in language. While language confirms communal identity and communication by converging the like-minded, different languages split humankind into distinct ethnic and racial groups. Nevertheless, language also has the capacity to bring together human ideas and concepts and to converge the mentality of all humanity by promoting good will.

On the other hand, religion often diverged into schisms, leading to the martyrdom of heretics and independent thinkers. Yet convergence was manifest whenever humankind worked in cooperation to found civilizations. Likewise across the world, ingenious problem-solving activated similar manual and mental skills.

Convergence also took place with the invention of writing systems as well as the translation of sacred Scripture and great works of imagination. Convergence was encouraged by the creation of centers of learning and libraries. Trade and global exploration also brought people together. From these experiences we learn that intelligence holds the greatest promise of ultimate human unity and integration.

Human ingenuity and conscience were also evident in systems of arithmetic and measurement. Calendars were invented which affirmed the shared memory of peoples. Laws, human rights and civilizations gave further evidence that symbiosis was characteristic of forms of human conscience. Each civilization demonstrated its invention and creativity but also left us proof in communal monuments of the mutual purpose of their communities.

On the other hand, past philosophies have illustrated human conscience. For instance, Plato, Aristotle, Stoicism and Epicureanism influenced much of European thought for more than two millennia. So too did religions such as Judaism, Christianity and Islam. Other religions as Taoism, Hinduism and world views as Confucianism and Buddhism also offered a range of life choices and destinies.

As a matter of fact, all religions seek to educate the human conscience. Monotheism itself matured human character. It guided us to consolidate our mental and spiritual energies; it nurtured a holistic vision of life. The concept of a Divinity urged individuals to integrate life's experiences into some final, moral meaning.

As human attributes, creativity and conscience should be the basis of a third millennium life philosophy. Both ancient Greece and Renaissance Europe embraced the ideal that life was for exploration, experimentation, invention and creativity. The aim was to realize one's aptitudes, talents and intelligence. This was their morphogenetic ideal of life. By contrast, in the Renaissance there was the contrary ideal of leading a life of religious devotion and obedience to one's Christian conscience. The individual was to shun sin and vices. This was the symbiotic ideal of life.

Juxtaposing the virtues of ingenuity and morality of these past civilizations, we discover how cultural and religious ideals express the essential morphogenetic and symbiotic nature of Humanity. Past cultures help us grasp the moral significance of our creative conscience.

Indeed, ideals organize and motivate men and women's lives. Moreover, humankind seems genetically creative and conscious of the need for compassion and mutual aid. Nature's Creative Conscience is not only adept at evolving the most effective plans and exquisite designs in all forms of life. Apparently, it also aims at evolving in humanity a moral nature that will be equally resourceful and skillful, yet just and humane.

Human perfection in its own right must involve creative and conscientious self-surpassing. We need both creativity and conscience to make sound life decisions. Self-realization is implicit in any definition of the moral life. Self-realization is a moral injunction to be obeyed by every individual. We evolve through pursuing some worthwhile purpose in life. We evolve by integrating the energies nature has so freely given us.

In contrast to the old reality of Darwin's theory of evolution, where nature was primarily represented as a "struggle for existence" and "the survival of the fittest" between species, twentieth century biology has provided us with a new and different reality. Ecology has presented evidence that nature manifests a great number of mutually beneficial relationships between distinct species, which have abetted their mutual survival.

On the other hand, in contrast to Darwin's injudicious interpretation of evolution in terms of mechanisms, twentieth century microbiology focused on the microworld of cellular transformation and integration. In other words, inner, somatic nature perpetually evolves ever more sophisticated means of survival.

Microbiology and ecology taught us that two universal processes pervade all life. Morphogenesis is the transformation and differentiation of tissues and organs. This microscopic process actualizes all stages of biological change.

Ecology discovered many examples of mutualism throughout nature. Correspondingly, we came to realize that the body of all living creatures is a totally integrated organism, which means that symbiosis is integral to all organic life.

In essence, morphogenesis and symbiosis present a new understanding of nature, a new reality.

Nature must now be understood from within the lifeform. To survive, an organism must deal with the limitations within itself. To survive, the individual must evolve its own inner strengths through creative self-transformation and symbiotic integration.

Morphogenetic intelligence in human beings is manifest by the way the individual uses ingenuity and deals with contingencies. Symbiotic intelligence is manifest by the individual's aptitude to coordinate and consolidate experiences, by the capacity to plan and design, by defining the meaning of one's experiences. It may be nature's way of urging humanity to understand the ultimate significance of human evolution.

The aim of *The Creative Conscience* is to motivate you, the reader, to draw on your own inborn resourcefulness and intelligence to create a life philosophy of your own.

PART I

THE EVOLUTION OF LIFE

1

DOES BIOGENETIC INTELLIGENCE ACCOUNT FOR EVOLUTION?

Twentieth century biology has certain reservations to Charles Darwin's key thesis in *The Origin of Species* (1859). However valid such objections to his nineteenth century theory may be, the intention of my references to it is not polemical. The main purpose of this chapter is to consider whether or not a biogenetic intelligence in nature provides a more viable explanation of the natural processes which actuate and actualize evolution. This chapter consists of four parts. Briefly it defines ecology and then describes microbiology. There follows an argument for the presence of intelligence in all forms of life. Last, the chapter undertakes to demonstrate how the processes of morphogenesis (nature's creative power) and symbiosis (nature's power of integration) manifest the primary activities of a biogenetic intelligence throughout Nature. Darwin's interpreters have not failed to point out that his theory of evolution is based on conjecture. Even so, it is conjecture founded on extensive knowledge, a superabundance of factual evidence and exact example. To do justice to this great scientist, let us quote Darwin's central thesis.

> As many more individuals of each species are born than can possibly survive; and as a consequence there is a frequently recurrent struggle for existence, it follows that any being, if it vary however slightly in any manner profitable to itself, under the complex and sometimes varying conditions of life, will have a better chance of surviving, and thus be naturally selected. Natural selection almost invariably causes much extinction of species, improved forms of life, and induces what I call Divergence of Character. (p. 68)[1]

(Divergence means the acquisition of dissimilar characters by related organisms in unlike environments.)

Not many biologists would question that there is a struggle for existence among competing species, or that natural selection can result in the extinction of competitors, or that the variation of life-forms improves their chances of survival. Yet the findings of twentieth century ecology and microbiology have put in stark relief the shortcomings of the nineteenth century interpretation.

Let us consider what we know from ecology and Darwin's own understanding of nature. Modern ecology investigates the interrelationships of organisms and their environment. Darwin himself was certainly clear about the intricate bonds throughout nature. Indeed, he stressed the fact that all species have "...infinitely complex relations with other organic beings and to external nature." Moreover, he admits "I...use the term 'struggle for existence' in a larger metaphorical sense to include dependance on one another." (p. 35) He concludes that "...plants and animals most remote in the scale of nature are bound together by a web of complex relations." (p. 36) Hence his theory foreshadowed the new science of ecology.

His admission seems to contest, to a degree, the thesis that natural selection almost invariably causes much extinction. Thus his "law of extinction" calls for modification.

Ecology provides the following qualifications. In the case where predator hunt prey to near extinction, predators gradually go hungry and starve because so little game remains . As a consequence, predators begin to die out and prey begin once more to multiply. As prey become more abundant, predators once more increase, and so the cycle repeats itself.

Or consider the scarcity of food. When this occurs, the females of many species do not produce young. In a similar vein, anthropologists report that food-deprived African women are unable to conceive. Hence this appears to be nature's way of "family planning." In other words, there is some form of intelligence in nature which regulates life and death relationships so that the Darwinian law of extinction is not an absolute.

Indeed, other processes in nature also circumvent and transcend extinction. For instance, there are examples of nature's ability to conserve life. Consider the dual functions of our nervous system. Faced with a life threatening situation, the sympathetic nervous system contracts the blood vessels to meet the challenge of the emergency whereas, when the danger has been dealt with, the parasympathetic nervous system dilates the blood vessels to return the body to its normal state. By this instinctive, neurological response, our body instantaneously "reaches a decision" which can determine whether we live or die. This is evidence of an ecological relationship between a creature's inner environment and the outer.

Another built-in response to our surroundings is that of the human eye, which reacts to light and dark in distinct ways. The pupil opens wide in obscurity to prepare us for the unexpected. In daylight it resumes its normal opening. The eye lens can also focus what is near and far, which enables us to foresee any danger or threat. Similarly our other senses serve to keep us from harm. Hence nature has provided us with sensory defense systems that help prevent our extinction. Prey are not entirely helpless or defenseless. Nature has seen to that.

There have been several theories explaining the extinction of species. One recent scientific theory refers to a natural disaster that occurred 65,000,000 years ago when a comet fell to earth near the Yucatan, Mexico. The impact raised such deadly pollution that it killed off dinosaurs worldwide. Another theory is that epidemics among predators and prey may have exterminated both. Other theories suggest that glaciation over millions of years had its effect. Or a plant blight defoliated whole areas of continents similar to the world's deserts, making them "lifeless" for millennia.

The fossil record of extinct species provides evidence of the earth's geological past. The record apparently substantiates the fact that various species have died out over the eons. What may have "caused" such extinction remains largely a mystery lost in time.

On the other hand, how is it the earth is still resplendent with life-forms? The reason might be that disasters sometimes act as a significant stimulus for the acceleration of organic regeneration and species transformation. Hence viable forms of life may have evolved due to tremendous bursts of animate life.

Darwin believed the cause of change lay in reproduction and inheritable variations rather than in spontaneous change in the adult organism. (p. 32) He would have modified his conviction somewhat, perhaps, in the light of the ecological discovery that the warmth of beach sand can determine the sex of turtle eggs laid in the sand. Hence mutability may be affected by the slightest changes in the conditions of the ecological environment.

On the other hand, there is evidence that the magnetites in rocks have changed direction, which seems to indicate that the earth's magnetic poles have changed over time. It is possible that magnetic deviations may have affected the biochemistry of animate life or electromagnetically influenced the genes of species. Perhaps the occasional meteoric collision affected the homeostatic stability of living cells in such wise as to initiate intrinsic mutations to strengthen themselves against the unexpected. Thus catastrophic collisions may not only have annihilated certain species but also have activated cellular experiments in others to create more effective, durable, and efficient modes of survival.

So ecologically speaking, species and the earth itself have coevolved over time. As the earth's environment changed, so too did plant and animal life. And the consequence of this coevolution is that the earth now is one integrated biosphere.

The topic of change naturally includes Darwin's conception of variation. Generally, the term means a divergence in an organism's qualities from those usual to its group. Darwin refines the generalization by stating "if it vary however slightly, in any manner profitable to itself, under the complex and varying conditions of life [it] will have a better chance of surviving, and thus be naturally selected." (p. 68)

The question is how and why such variation comes to pass, or how such selection takes place. Our human experience tells us variation must come about by some inner need or want in the organism. Creatures change out of necessity or curiosity. Variation indicates an animal's response to change in the environment, whether favorable or adverse. Then again the organism also responds to inner urges, impulses, hunger, thirst or drives. Ultimately, variation may be due to some form of cellular or somatic intelligence, which informs the organism of its homeostatic needs or stimulates it to grow and transform in a certain way. This change usually starts as a cautious experiment, which gives the life-form a new source of energy. For instance, it is probable this took place when the first plants and flowers learned to follow the sun to absorb its energy.

From the mollusc to the jellyfish to the mammal, from the invertebrate to the vertebrate, life-forms responded to their inner needs and to the outer environment. A change in one initiated a change in the other. In a broad sense, there exists a homeostatic bond between the individual life-form and the nurturing environment it lives in. (Generally, this is what is studied by ecology.)

Of course, Darwin was well aware of the power of change in nature. He repeatedly expressed wonder at the plasticity of nature in its capacity to reorganize its structure and constitution. (p. 75)

Today's microbiologist would regard such mutability as energized and designed from within.

The science of microbiology teaches us to see the minutest forms of life. Without a microscope, germs, cells and their structures remain invisible. Without a high powered microscope, a vast number of microorganisms remain indistinguishable. Such infinitely fine specks of life make it clear that bioprocesses are at work throughout living nature which abet their own laws and influence our existence.

In the microbiological realm, cells reveal considerable intrinsic activity at the same time they are able to enter into cellular combinations that become multicellular organs of all sorts. Microscopically, we witness their power of generation and of consolidation into more complex and intricate forms of life. They are able to transform and integrate into entire life-support systems, which, in turn, become evolved life-forms.

Everyone knows about metamorphosis. It is the outer transformation we observe when the form and structure of a caterpillar becomes a butterfly or a tadpole becomes a frog. Few of us think about the inner cellular processes responsible for the creature's genesis from the earlier to the later form of life. Similarly, variance may be considered the result of nature's creative experimentation with its own bioprocesses.

We stand in awe at nature's inherent power to change and transform. The microbiologist studies nature's plasticity in the cellular, genetic and somatic spheres, which activate corresponding changes in structure and constitution of living things. We learn from microbiology that variation and change are initiated to meet inner needs as well as to effect a more successful, resourceful relationship to the outer world.

Although these bioactivities are invisible to the naked eye, they can be perceived through the electronic microscope. Are we supposed to believe such interactivity is not somehow "sensed" at the cellular, somatic or infraconscious level of "self" awareness? As animals, sometimes we ourselves become conscious of such activity at the neurological, glandular, physiological, emotional and cerebral level. It would be astonishing if there were not some measure of "consciousness" or "feeling" in the tiniest living things.

In a way, the cells of our body seem able to germinate as do seeds. (Of course, plants also have cells, each with a wall, plasma membrane, mitochondria, vacuole, cytoplasm, nucleus and the like.) The single animal cell, interacting with other cells, is capable of performing fundamental functions that transmit life and make structural units which create independent beings capable of reproducing their own kind.

As microactivities, our cells and organism seem to incarnate modes of intelligence and purpose. Our neurological network, body and brain not only respond instinctively to any threat from the environment but also direct us to pursue and fulfill our inherent needs.

Proof of such intelligent microactivity is the way the immune system responds automatically to a wound or to microbes invading the body and causing sickness. It rushes immunoglobulins in response to specific antigens to counteract the invaders by neutralizing their toxins and harmful bacteria. The immune system's counterattack is initiated by the generative morphogenesis of the body. On the other hand, symbiosis actually "heals wounds" and "cures" sickness by reintegrating the traumatized tissue or by restoring infected cells to health. Hence it is cellular intelligence that fine tunes the body's life-defense systems to keep us alive.

Microbiology has also made us aware of genes and their role in life. A gene is described as a sequence of nucleotides in DNA and RNA that is located in germ plasma. It functions in controlling and guiding inheritable tendencies in all living plants and animals. Moreover, there are gene complexes which act in mutually cooperative ways. Periodically and alternately, they seem to maintain control or actuate change. Sometimes they act to bring about meaningful growth or metamorphosis. At other times, they act to maintain healthy homeostasis. As microactivities it would seem our cells and genes incarnate modes of intelligence and purpose.

In fact, Robert Wesson in *Beyond Natural Selection* (1991) observes that "Some biologists exalt the gene over the organism and demote the animal itself into being merely the means of replicating genes." (Dawkins,1976) In the neoDarwinian perspective, evolution is the result of competition among genes, duplicating genetically the struggle for survival of the fittest." (Wesson p. 11)[2]

When we examine how a single cell organizes into tissues with specific functions, these tissues into organs, and these, in turn, consolidating into systems, we realize that cells and genes have acquired through evolution the power to perform specialized functions in life-forms of all kinds. If we consider DNA replication (mitosis) and sex cell division (meiosis) surely these microactivities are governed and guided by a life energy which manifests itself through modes of intelligence. Not only is its objective to stay alive and to reproduce itself. Every stage of an evolved individual's genesis and of his species evolution represents a stage of maturing or evolving intelligence.

Before presenting further evidence of intelligence in nature, it is appropriate to consider evidence against the argument for it. Wesson points out that "An unconditional thesis of neoDarwinism is that "Chance is the source of true novelty." (Crick, 1981,58) Wesson concludes from Crick's thesis that innovation in evolution can only arise "from errors in our reproductive process." (p. 9)

Crick's neoDarwinism is difficult to accept. Evidently he assumes that life-forms are mindless, basically incapable of surviving by cellular-organic intelligence or by learning from life and death experiences, which would teach them to survive. Yet, to the contrary, the daily challenges of staying alive must, occasionally, have made the most intrepid or curious creature aware of never before tried possibilities of surviving. While chance may well have played a role, it is more likely that necessity made them alter habits or strategies, change their habitats, or vary their activities. Lessons were learned or resulted in death. In those who did survive, those lessons must have become instinctive by the close observation of a parent, by mimicry, by learning on their own. There was punishment for mindlessness and reward for innovative thought and foresight.

Chance can mean luck, contingency, or opportunity. If chance did play a role in evolutionary "novelty," probably all three conditions had their influence on the outcome.

We do not have to accept the neoDarwinist determinism that chance seems to embrace. Consider the wide variety of life-forms and the various organic systems that sustain them. Consider the societies formed by insect and animal. Consider the ecology of our intricate biosphere. In the face of such evidence, it seems quite fantastic to assert that the lives of life-forms are determined by chance alone, especially when we bear in mind the survival success of so many species over millions of years.

Moreover, to struggle is to seek, to attempt, to strive, to undertake. Every individual and every species must have listened to an instinct, an inner call, an inner fear, a purpose to stay alive yet another day. Surely desire, hunger, thirst, and need bespeak necessity, rather than chance, as the driving force behind change.

The biosphere of the earth with all its intricate symbioses has not left the fate of the individual or the species to chance. To the contrary, life as nature and nature as life seems to have nurtured in every living organism the capacity to become stronger, more resourceful, more circumspect, more reflective, in sum, wiser; for to fail to come to terms with reality meant self extinction. The life instinct throughout nature has confronted death with the resolution to live yet another day. In all forms of life, intelligence was at work.

What we know about all species on earth allows us to suggest a meaningful interpretation of evolution. Those that have survived to this day have done so by adroit adaptation to their special environments, and the adaptation itself is sufficient evidence that all life-forms use some mode of intelligence to survive. Intelligence is active throughout nature, and its presence accounts for the processes which bring about morphological change.

Neither "external" environmental conditions alone nor "internal" organic conditions alone can explain stability or change in the structure or function of any organism. Rather, a life-form is the result of inner symbiosis, which establishes its own intrinsic homeostasis. On the other hand, the life-form continues to exist primarily through its own sensible, symbiotic relationship with both the abiotic and biotic environments.

In contradistinction to any mechanistic interpretation of nature's processes or evolution, the intelligence of a life-form is not only in its "parts" but in its very substance. Living cells evince intelligence by their capacity to differentiate and specialize.

Any analogy between man-made mechanisms or electronic devices and the processes taking place in a life-form is merely a mirage image of the truth. We ourselves are not the automatons that the French philosopher René Descartes (1596–1650) thought us to be. Life is in us and we are in life. We illustrate the significance of all life-forms that went before us. Intelligence, resourcefulness, and the instinct for survival have led us to sense that life has some meaning, and it is up to the individual to discover what it is.

Surely in some manner and to some degree, all life-forms on earth must have felt in their cells and tissues there was some reason for being alive. And evolution most assuredly came about by animate means, natural processes and an endless series of partially perfected designs suited to the ever changing environment of the earth.

Perhaps it was sunlight that awakened life in the sea and on land. Once aroused, countless cells evolved and united, and in this activity discovered they were alive and began to direct their own existence as best they could. With that awakening, intelligence appeared in the universe.

It is a remarkable fact that life seems to create energy similar to the way the sun creates light and heat. In its most obvious form, sunlight stimulates photosynthesis in plants which enables the plant to create forms of energy. It needs to grow, to defend itself and to reproduce. Hence the plant is able to adapt to the condition of the environment not only to survive and use it but also to multiply so that its species might continue to live for thousands of generations. Surely even the most conservative biologist would concede that the ability to survive manifests some form of intelligence.

It is understandable that some biologists question the implication that nature has a will of its own. Yet a number may be prepared to accept the evidence that intelligence in nature seems to have influenced the survival of the species.

In contrast to Darwin's own assertion that evolution probably proceeded by chance mutations, the use of the expression *experimental mutations* might be more appropriate to explain evolution in terms of ecology and microbiology.

There is substantial evidence of some inveterate form of intelligence in nature. Everywhere, we perceive plants and animals not only adapting skillfully to their environment but also through autoinnovation inventing defensive strategies of all kinds. Indeed, to a degree, both plants and animals are able to manipulate their surroundings, to find new sources of food, and to ward off encroachers and predators.

Furthermore, there is sufficient evidence at the cellular level of life that the individual organism can defend itself against noxious invaders and that it can ensure its own optimum health and survival. Moreover, there is an abundance of micro and macroevidence to infer justly that some universal process in nature maintains the homeostatic harmony of all living cells, and yet it is equally capable of actuating the dynamic development of every plant and animal alive.

Finally, that same intelligence in nature very probably actualizes evolution over endless eons of time. What we perceive in the regularity, rhythm, pattern and cycles of nature—all that is perfected form and consummated design in the service of the individual and the species—makes it sufficiently clear that some form of active Intelligence permeates and animates all Nature.

As we have seen, contemporary scientists have studied scrupulously the microbiology of processes believed to animate the smallest life-forms known to humankind. In turn, ecological studies have revealed that these selfsame processes are present throughout the earth's entire biosphere.

If nature can be defined as "some creative and controlling force in the universe," the natural sciences have found two processes present in every form of life.

Microbiology interprets one ubiquitous process as morphogenesis. This process is active not only at the cellular level of life but also throughout all living organisms animal and man. The scientific definition of morphogenesis states it is "...the origin of evolution of morphological characteristics; the growth and differentiation of cells and tissues during development." (Barnhart, 417)

The complementary process is symbiosis. Commonly it is defined as "the intimate living together of two dissimilar organisms in a mutually beneficial relationship, especially mutualism." (Tenth, 1194) However, within any living thing, there also exist relationships which, in effect, constitute a symbiosis among specialized cells that form the organism. At the microbiological level, symbiosis accounts for the coalescing of tissue, muscle, bone, various organs and life-support systems.

The purpose of *The Creative Conscience* is to interpret the implications of these discoveries and to extrapolate the lifelong meaning they have for the individual. The intention is to help readers become aware of how these processes can significantly affect their lives.

Further exploration of these two bioprocesses reveals how far they permeate life.

Morphogenesis acts via a two-phase process, which may take place concurrently or periodically. One phase is to direct the life-form's attention to explore the environment surrounding it. The purpose of this exploration is to seek nutrients, the useful, the life enhancing. (In plant life, this phase corresponds to the extension outward of the roots, stem and leaves.) Thus an organism initiates a search of its environment to find sustenance not only to survive but also to reproduce its own kind. To be sure, selectivity is used to distinguish the harmful from the helpful, the toxic from the nourishing.

In addition, it has a corollary phase, which turns inward to explore the limits and nature of the life-form itself. This phase helps develop cellular functions or organic structure within. Hence this phase enables the organism to transform the raw materials, absorbed from the outside, into useful nutrition and to build or repair whatever elements are in need. Correspondingly, it examines its own activities and discards less viable strategies of survival for those more viable.

Thus morphogenesis itself searches the environment within and without both to sustain any established homeostasis within and to engender more effective, durable means of survival. Hence in itself, morphogenesis is an interacting, dialectical process which activates and actualizes whatever is required to maintain optimum health and life itself.

Beyond its own intraactivity, morphogenesis interacts with symbiosis. In an organism, the primary task of symbiosis is to conserve the energy within by cementing together tissues and strands into shapes and networks which grow increasingly intricate and efficient. In this manner, symbiosis coalesces, consolidates, and integrates the substance and sense of the biodata gleaned by morphogenesis. Moreover, by so doing, symbiosis becomes ever more aware of itself as a process. The osmotic, metabolic, and organic processes of life evolve into more complete, more complex forms. As they develop throughout a life time and evolve over eons, such forms by their characteristic conservation of energy and organic intricacy develop consciousness and memory.

By working synergistically with morphogenesis, symbiosis enables the life-form to evolve more knowledge of life itself. However, eventually symbiosis interacts with itself. It becomes conscious of a purpose. This is to govern and guide the curiosity and creativity inherent in morphogenesis by "teaching" it not only to accept its own limits but also to build on its own experiences with life. Thus in a figurative sense, symbiosis gradually evolves a "conscience" in the individual life-form, which, over great spans of time may evolve in a few species an instinctual "wisdom" as to how life is to be lived.

Twentieth century biologists argue with Darwin's belief that evolutionary change came about by gradual variations. Stephen Gould and Niles Eldridge maintain that a dichotomy exists between the stasis of a species and their (periodic) rapid change. (Stanley, 1979, 13–22) quoted by Wesson, p. 14.

A different explanation of evolutionary change seems more probable. Under favorable conditions, an organism's transformation may be accelerated. Such a period would be sensed as an opportunity to extend its domain or to evolve further. In turn, growth would likely be followed by a time of symbiotic consolidation and integration.

On the other hand, unfavorable conditions (as severe seasonal drought, climate changes, natural disasters) may produce a contrary effect. As during seasons of hunger, the animal body automatically shuts down the process of starvation so that the individual and species can survive.

In other words, in as much as favorable and adverse conditions of life recur, an organism would eventually learn to develop opposite but complementary capacities of survival. An example of this capacity is the endocrine system which regulates stress, metabolism, blood pressure, and immune response.

Thus if biogenetic intelligence exists in nature, when adverse changes in the environment occur, morphogenesis initiates appropriate defense measures, and symbiosis consolidates those measures to counter and overcome life-threatening events.

Current microbiological speculation also considers the possibility of the genetic initiation of evolution. Wesson observes, "Evolutionists differ in their emphasis on single genes or combinations...If most traits are the effect of multiple genes, regulatory genes, or gene enzyme systems, large changes become more conceivable as results of recombinations." Although regulatory genes are poorly understood, they seem to turn batteries of genes on and off. (Wesson, p.15)

This regulatory capacity of certain, specialized genes may be understood biochemically. The dearth of any chemical, vitamin, mineral, enzyme or other bio-substance may trigger an on-switch to initiate a search for what is needed; or vice versa, a surplus or surfeit may trigger an off-switch to shut off the excess. This capacity is easily characterized by our responses to hunger and thirst and to other biological or even psychic needs. As such, they prompt decisions and actions to fulfill the want or to prevent an overabundance from harming us. This simple fact proves how our biological nature has taught us the necessity for decisions and appropriate action. The regulatory genes may be responsible for initiating periods of morphogenetic and symbiotic activity.

Thus far the discussion has underscored the thesis that the evolution of organisms derives from the interaction of morphogenesis and symbiosis. The general purpose of morphogenesis is to discover more sources of life-giving nourishment and more efficient ways of surviving. On the other hand, the role of symbiosis is to ensure the consummation of successive orders of organization in life-forms. Wherever symbiosis integrates the needs of morphogenesis, there you find nature's plan for perpetuating life. The dialectic of morphogenesis and symbiosis is what actuates and actualizes evolution.

In an attempt to update Darwinism, the neoDarwinists sought to synthesize Darwin's natural selection with Mendel's genetics. Although at first its seemed a satisfactory synthesis, it soon appeared outmoded. "Biology increasingly must deal with process and pattern, with self-organizing and self-regulating systems." (Wesson, p. 36) The above description of morphogenesis and symbiosis demonstrates the process and patterns of such systems.

Wesson also notes certain salient characteristics of morphogenesis. The essence of the process is "symmetry breaking." (p. 36) This assertion corroborates the understanding that the process pursues growth, extension and exploration of both the outer and inner environments. It has explorative, experimental, and creative characteristics.

Moreover, Wesson makes clear the significance of the process in the widest terms possible. "Animals respond to external conditions according to inner dynamics. Self organization is the essence of the origin of life and its complexification, that is, evolution. It lies at the heart of morphogenesis, ecology, and the aggregation of human culture." (Wesson, p. 36)

Much of Wesson's *Beyond Natural Selection* offers serious reservations to some of the key concepts of Darwin's theory. Although Wesson's own examples point to nature's imperfections and the incomplete development of species, these anomalies also suggest that life is still in the process of making order out of the original chaos. Life may still be in the process of working out perfections. Or it may be coming to terms with the limits of cellular organic perfectability. In any case, nature still seems to be coping with the designs of its creativeness.

A key idea in evolution is mutability. Regardless of the circumstances and conditions which stimulate change, it is immanent in life itself. In the context of microbiology, it is apparent that such mutability is energized and designed from within. As a result of an organism's inborn instinct for exploration, experimentation and creativity, its inherent genesis actuates growth and transformation. Yet such actualization is not simply a matter of mindless proliferation. Morphogenesis is monitored by symbiosis. The task of symbiosis is to coalesce these energies, to concentrate and integrate them into viable mutations that can endure.

Darwin alludes to the web of complex relations among distinct species. Obviously such give and take relationships require reciprocity, which calls for morphogenetic adjustment and symbiotic habituation. Moreover, such interacting processes not only establish a web of life but also make possible the mutual evolution of species over time.

In more advanced species, the evolution of perception and mental skills is a corollary manifestation of the way they surmount the limits imposed by the environment. Moreover, the evolution of organic, neural, cardiovascular, respiratory, reproductive, muscular and skeletal strengths is positive evidence of some ageless intelligence animating nature. This intelligence most likely arose out of the cooperation of the two predominant bioprocesses we have been discussing at length.

Mutual interaction has other implications. Conditions of the environment or limitations in the organism itself can trigger a response to an outer challenge or inner need. When stimulated to explore any difficulty or when motivated to overcome its own biolimitations, a life-form's response would indicate the emergence of an instinctive intelligence.

Furthermore, implicit in Darwin's natural selection, there is evidence of a creative intelligence, whose purpose is to experiment with life's alternatives and options. Indeed, this is the potential and responsibility of morphogenetic intelligence.

Complementing this process is the pervasive purpose of symbiotic intelligence, which is to monitor the environment, the threat it poses, and the opportunities it affords the organism. Within a life-form, symbiosis coalesces, coordinates and unites everything into an intelligence capable of survival. In sum, the morphogenetic intelligence a human acquires by experimenting with life's opportunities is monitored by the experience our symbiotic intelligence has accrued in a lifetime.

The biological correspondence between Nature and human nature is unmistakable. By our morphogenetic curiosity, exploration and experimentation, we satisfy the deeply sensed need for creativity in our lives. By our symbiotic concentration, self-discipline and consolidation, we gratify the other need of humanity to integrate our experiences by pursuing a purpose in life. Through listening to the two voices of our nature, we begin to understand the meaning of our human destiny.

Thus, symbiosis interacts with morphogenesis in Nature and in human nature. Together they drive and shape our somatic, emotional, rational and creative essence as an evolved species. Together they interact to effect the total entity that makes up our living being. Together, they create the conscience of our species. They account in large part for our instinctual knowledge and wisdom.

In the context of human evolution, the morphogenetic intelligence acquired by our exploration of life's opportunities is guided by the memory of symbiotic intelligence, which has gleaned the lessons of a lifetime. In this sense, symbiotic intelligence is synonymous with humankind's nature-born conscience. Finally, the interaction of morphogenesis and symbiosis accounts for the creative conscience endowed by Nature itself to humanity.

The biologic we now perceive in nature surpasses much scientific and philosophical reasoning that we had hitherto considered totally valid and viable. Not only do we need to reevaluate our ways of thinking in the fact of Nature's intrinsic creativity and self-organization. Over millions of years, morphogenesis and symbiosis have been involved in the evolution of species. The processes of nature discovered since Darwin's *Origin of Species* suggest there are forms of cognition we yet have to learn, if we are to further our own remarkable evolution.[3]

In sum, biogenetic intelligence accounts for life and for human evolution.

PART II

———

A 2,500 YEAR HISTORY OF BEING AND BECOMING

2

BEING AND BECOMING

Over the past 2,500 years in Europe, thinkers have undertaken to describe three principles that were believed to explain the powers governing the universe. Professor Arthur 0. Lovejoy's *The Great Chain of Being* (1936) retraces the history of these germinal ideas which guided the thought of philosophers, theologians, and imaginative individuals.[1] He defined the unit ideas as *plenitude, continuity* and *gradation*, all of which characterized the archetypal processes, Being and Becoming, as the ultimate powers energizing existence.

Among ancient philosophers were Plato, Aristotle, and the Neoplatonists. Theologian philosophers were the PseudoDionysius, Augustine, Aquinas and Averroës. Among the scientific-minded were Aristotle, Giordano Bruno, and Robinet. The most imaginative were Dante, Leibniz, Schleiermacher, and Friedrich Schiller. All these thinkers embrace one or more of the three principles as providing the essential explanation of the spirit and substance of the universe.

Rather than emphasize the academic essence of each thinker's philosophy, this chapter aims at pursuing a more down to earth purpose: (1) not only to examine the value of the three principles to our living life creatively, sensibly and morally, (2) but also to explain how morphogenesis and symbiosis furnish scientifically verifiable interpretations of life's universal Being and Becoming.

The Principle of Plenitude

Since antiquity, Western man has undertaken to make the world appear rational. In *Timaeus*, the philosopher Plato argued that the created world was self-sufficient, so designed that it initiated all its own activities and processes. To this vision of Self-Sufficient Perfection was added the concept that the universe is Self-Transcending. This explained both its unity and the multiplicity and variety of its life-forms. Because every conceivable form of life was present, Plato named the universe a *plenum formarum* where each form could realize its own potentiality of being.

Of special interest is the assertion that the universe is perfect and self-sufficient. It is meant to humble man as to his status in existence. Without and before man, life maintained itself. In other words, life as a phenomenon manifested a truth greater than man himself.

In subsequent centuries, Neoplatonism elaborated on Plato's theory of emanationism in the *Timaeus*. This theory came to mean that through intermediate stages emitted in a series of emanating radiations from the Godhead, the world came into being. Thus God's presence in the universe could explain the self-transcending power of life itself.

In early Christianity, this vision of God's powers was reinterpreted. For instance, the Athenian disciple of Paul, PseudoDionysius, defined the divine attributes to be "love"and "goodness." By this he meant "the immeasurable and inexhaustible productive energy in the universe." Similarly, late Medieval writers found that God's love "consisted in the creative and generative rather than in the redemptive or providential office of the deity." (p. 67)

On the other hand, in fourth century Christendom, Augustine stated all things in the universe should be. For this reason it was rational. This was what plenitude meant.

In the thirteenth century, Thomas Aquinas asserted that the orderly perfection of the universe affirmed the principle of plenitude. Moreover, "the universe is its own reason for being" and is not "merely a means to man's salvation." (p. 77)

Aquinas's statement comes very close to modern man's understanding of existence. The world and everything in it is clearly its own reason for being. In other words, if the individual wants to ensure or realize "salvation," that person needs to discover a reason for being in this world.

However, not only were Christians preoccupied with discovering the sense and significance of existence. Both Moslem and Jewish medieval thinkers wrote on the principle of plenitude. For instance, the Spanish-Arab philosopher Averroës saw the existence of species as based on God's having created all life-forms. God's presence assured that the principle of perfection and completion prevailed in the world.

Although later thinkers expressed their own versions of the principle of plenitude, we need here to draw some conclusions about the philosophical significance of the principle. It should be noted that after Plato and the Neoplatonists, most thinkers declared faith in the rationality of existence, in spite of its contingencies and irrationalities. Once the fundamental faculty of the mind was assumed to be Reason, these thinkers framed all knowledge and experience into a rational paradigm. As already made evident, though they were conscious of a generative power in the universe, they emphasized the rationality of the world.

The last example of the principle comes from the sixteenth century Italian philosopher Giordano Bruno. Inspired in part by the theory of Copernicus, Bruno conceived the universe as populated by an infinite number of stellar systems. Thus the ancient argument for plenitude and sufficient reason led him in his *De Immenso* (1586) to argue that "the divine essence is infinite throughout existence." (pp. 116–117)

The Principle of Continuity

Aristotle was "responsible for the introduction of the principle of continuity into natural history." (p. 56) Although he recognized there could be multiple systems of natural classification, he suggested that all animals might be arranged in a natural scale, according to their degree of "perfection." (p. 58) In his *De Anima*, he proposed a hierarchy of organisms, based on "powers of soul." Hence each higher order possessed all the powers of those below it in the scale. (pp. 589) However, there existed organisms which have not yet realized their potential, and there were others at superior levels of being. (p. 59)

Two millennia later, the eighteenth century philosopher J.B. Robinet questioned the value of classification. Indeed, he contended it was a plain denial of nature's evident continuity. He asked: What continuity can there be distinguishing the organic and inorganic, the animate and inanimate, the rational and the nonrational? Classifying creatures had little to do with life's continuity. Rather, there must once have been a prototype for all plants and properties. However, for Robinet, the prototype was an intellectual principle which would realize itself in nature. (p. 229)

Moreover, in nature Robinet discerned movement in one general direction, "a striving toward a particular goal via trial and error toward a consummation not clearly foreseen." (p. 280) He also stressed the fundamental reality of nature was activity, not matter.[2] In fact, activity was the universal essence of being (*le fond de l'être*), a process not yet finished. (p. 282) Later thinkers called Robinet a "forerunner of evolutionism." (p. 286)

The German philosopher Immanuel Kant also considered the Chain of Being as a strict continuity. (p. 241)

The Principle of Gradation

Aristotle seems to be the source of the third principle inherent in nature. He encouraged two diametrically opposed sources of conscious and unconscious logic. One mental habit is to think in well-defined, discrete, class concepts. The other is inherited directly from Nature, which "refuses to conform to our craving for clear lines of demarcation." Indeed, he argued that the whole notion of species seems an artifice of thought with little to do with nature's fluidity." (p. 57)

In fact, based on the perception of our inborn powers, Aristotle's *De Anima* suggested a hierarchical order of organisms from man's own rational nature to levels of being superior to man. (pp. 589) From this notion of an ontological scale in nature, combined with a hierarchy of intelligence, Professor Lovejoy himself identifies this principle as unilinear gradation. It should be added to the other two principles. (p. 59)

Although inspired by Aristotle, it was the Neoplatonic theory of emanationism that fused the principle of plenitude with the principles of continuity and gradation. Envisioned as a self-transcending dialectic, the "scale of being" became the essential conception of Neoplatonic cosmology. "All things were linked in a continuous succession downward from the Supreme Mind, filling them with life down through the lesser beings of existence." (p. 63)

As a consequence, during the Medieval period, "the primitive Christian conception of a loving Father in Heaven" was converted to a conception of a Godhead from which all earthly life emanated. (p. 68) In other words, nature was a process emanating from the Godhead, which gave life to all creatures. As such, an omnipotent spirit, intelligence or life force was responsible for all change in Nature.

The fourteenth century poet Alighieri Dante echoed this intuition of gradation in nature. In his *Paradiso* he expressed the belief that divine goodness not only created mortal and spiritual beings but also, above them, immortal beings as well. It was God's infinite goodness that made certain of their realization. (p. 69)

During the Renaissance, in biology the principle of continuity and gradation played a decisive role in the history of being and becoming. The authoritative principle of the period became the conviction that all living beings were linked to one another by graduated affinities.

The Meaning of Nature's Principles

With these representative examples of the three principles thought to account for life in the world, the reader may well ask "What do these principles have to do with the way I live my life?"

The answer is within easy reach. Plenitude in nature offers an obvious lesson for everyone. The lesson is that, above and beyond biological gratifications, each of us feels the need to live a range and variety of life experiences. As Helen Keller aptly said, "Life is an adventure, or it is nothing ."

Any literate person with skills and education can indulge this need in endless ways: hobbies, handicrafts, apprenticeship, sports, the arts, music, museums, reading, writing, learning about one's own culture, travel to foreign lands. In fact, never before in human history has there been such a plenitude of decent pleasures for the common man and woman. Life is not only for the well-to-do.

Beyond these many opportunities to gratify one's appetites and curiosity, the second lesson of nature brings us to muse about the sense of our private lives.

Each of us needs to discover the continuity of our existence. While our literacy, special skills and education enable us to meet our vocational responsibilities and practical obligations, while material success in life assures us that we can know plenitude, something more is required to make us fully human. If we wish to learn the sense of our own lives, we are obliged to come to terms with our life experiences. Some experiences have given us rewarding moments. What did they mean? If we seek out the continuity which reveals our basic character through succeeding identities at various stages of life, we can discover the meaning of our destiny.

Perhaps the search for continuity is one reason that stories have such compelling fascination for us. They may sum up fictional experiences, but they provide us with insights into life's real significance. Sometimes in our own lives, we become conscious that episodes or events make good stories. At times, certain unforgettable experiences seem to have shaped our lives. If we remember some of the great stories and poems we have read, we realize they have shown us the varied destinies of humanity. They etch our hearts and minds with wisdom.

Nature has a third lesson. If linear gradation manifests itself most obviously in the evolution of earth's life-forms , the principle also invites us to consider how it might apply to our simple lives. To be sure, an individual's life-long changes are a self-evident clue to how the principle of gradation applies to our lives. Clearly we pass through successive stages: babyhood, childhood, adolescence, young adulthood, full maturity, middle age, and advanced age. How are we to understand each? We remember how we assumed different roles and identities during our lifetime. What we need to learn is that each stage has its own purpose. In point of fact, no age is better than another. The greater the number of years with a sound mind and sound body, the longer the perspective and the clearer the vision of time and its meaning.

Not only that. We do not live lives devoid of moral responsibility. The more we understand the importance of nature's principles of plenitude, continuity and gradation, the better prepared we are to educate our offspring and descendants. We owe them our knowledge of nature and wisdom.

With self-knowledge, insight into time, and thoughtfulness toward our own kind, we begin to appreciate our common destiny with the rest of humanity.

Toward a Comprehension of Nature and Evolution

To better understand how the three principles evolved in the minds of later thinkers and became integrated into an advanced intuition of the nature of the universe, we need to consider the seventeenth century German philosopher G.W. Leibniz. While conceiving the three principles to be the essential characteristics of existence, his theory of monads made an important addition to the concepts.

For Leibniz the "chain of being" was made up of the totality of the world's monads. (For Leibniz, a monad was an elementary spiritual force from which evolved all material properties.) These monads represented a "hierarchical sequence from God to the lowest grade of sentient life." Moreover, they characterized levels of consciousness, which gave the universe its adequacy and clarity." (p. 144) In other words, finite things were part of God's rational order." (p. 166)

Hence Leibniz held two distinct visions of the universe, the first based on the eternal idea of the Divine Reason. That order created the essential structure and immutability of the world. (pp. 161–2) In the second vision, he conceived time as a continuous augmentation of realized values. As the most significant aspect of reality, change became the most indispensable mark of excellence because it led to perfection. (p. 162)

Furthermore, he came to believe that "all forms [in nature] are realized in the order of time." In fact, "...even the species of animals have been many times transformed." Thus, his own understanding of nature evolved. If earlier he declared "nature is always perfect," later he saw it as "always increasing in perfection."

Being and Becoming

Leibniz, his contemporaries and successors modified their natural theology and metaphysics. They gradually converted the once immutable "Chain of Being' into endless Becoming." (p. 259)

The marked change in Leibniz's understanding enabled him eventually to give us the most persuasive conception of the Chain of Being. He now considered that the monads emanated in a hierarchical sequence from God on high to the lowest form of life. All these orders of natural monads formed a single, continuous chain of being. (pp. 145)

Of special interest is that Leibniz conceived these successive beings as manifestations of consciousness.(p.144) Because these beings display "different degrees of perfection, some essences have a greater claim...to existence than others." Thus each being...would be measured by the rank, or degrees of (its) excellence." (p.178) From the modern perspective "every being would have the right to aspire to existence in proportion to the amount of perfection it contains...." (p.178)

The reader may wonder where he or she would be in humanity's scale of being. The level of achievement might be gauged by past and present successes. Or one's lifetime creativity and moral commitments could add up to a final sum worthy of being human. Put in terms of our thesis, morphogenetic activity and accomplishment might provide a scale for conscious self-evaluation. For instance, the scale would judge a person for the exploration of talents, innovations in one's job, the use of imagination in daily living, or even trying out different life styles. In other words, evaluating an individual's resourcefulness or independent thinking may be one way to measure a person's curiosity, creative intelligence, and capacity for adventure.

On the other hand, symbiotic activity and attainment may be a complementary scale of conscience. For instance, such a scale might determine an individual's skill in coordinating and subordinating activities aimed at achieving goals. It might examine the self-discipline in concentrating for long periods to advance or complete projects. It may be used to assess the perfection of one's designs. It might test the capacity to bring one's experiences together into sensible meanings. It might judge how well one can integrate one's emotional needs and intelligent pursuits so as to gain a measure of mastery over one's life.

Such conjectures lead us to admit that Leibniz's principle of gradation, based on a graded scale of monads, with God at the summit, offers a truly novel idea. With monads at different "degrees of perfection," we reckon they must attain distinct degrees of excellence. Since the "higher" can lay greater claim to survival, Leibniz's scale of gradation offers a fresh intuition as to the meaning of human evolution.

That meaning would certainly have to do with degrees of accomplishment that an individual has achieved through the use of natural intelligence. On the other hand, humankind has also developed various forms and degrees of moral intelligence.

If the former includes resourcefulness, curiosity, ingenuity and the ability to survive one or more environments, the ethical form of intelligence is more comprehensive in meaning. Moral intelligence includes the ability to concentrate and complete tasks. It demonstrates the power to exercise self-discipline. Yet moral intelligence is more than self command and self mastery. It also manifests humankind's conscience through empathy for all living creatures and compassion for the spiritual and survival needs of humanity. These characteristics would also mark one's place on the scale of Homo sapiens moral evolution.

One last comment on Leibniz is worth making. He maintained that everything has some reason for its existence. Each "…is logically grounded in something else which is logically ultimate." (p. 146) Today we feel that every living creature has a reason for being. Indeed, in as much as humans evolved from Nature, we too must have some reason for being, for our species emerged from the universal bioprocesses of nature.

The Value of the History of Ideas

Lovejoy's *Great Chain of Being* is an exceptional example of a sophisticated, scholarly use of the history of ideas. In his preface, he defined the "chain of being as the descriptive name for the universe." (vii) Throughout much of European history, the metaphor was used "…to show that the scheme of things is an intelligible and rational one" and to describe "the structure of nature." (p. viii) However, Lovejoy admits that "the concept was not a generalization derived from experience nor easy to reconcile with the known facts of nature."(p. 188)

As already seen in this chapter, the apparently apt metaphor of the structure of the rational universe not only changed in successive centuries, but also it reversed its meaning from static being to dynamic becoming. This fact necessitates a caveat against the danger of inadequate interpretations, whether practiced by critics, scholars or scientists.

It is understandable that the general, intellectual, moral and cultural climate of an era (its Zeitgeist) is bound to be captivated by ideas, analogies, and metaphors which seem to sum up an entire world view. However, caution is recommended about the assumptions of our own generation and of our own period of history. Sometimes our most fundamental assumptions, derived from science, religion or populist philosophy, may be patently false, astygmatic, unsound, or emotionally distorted as in the case of fanaticism and fundamentalism.

Lovejoy explains his method. The historian of ideas "investigates the persistent dynamic factors and ideas that produce effects in the history of thought." (p. 5) He does this by retracing "unit ideas," which are "component elements of systems and philosophical doctrines." (p. 3) "The historian seeks out the disposition to think in terms of particular types of imagery."(p. 7) He isolates any unit idea which figures in any important degree whether in philosophy, science, literature, art, religion or politics "to elicit the role of a specific thesis or argument in history." (p. 15)

These statements require some comment. In as much as human biology has a great deal to do with the way we imagine and think, a better term might be proposed to discuss the mental associations an image stimulates and the concepts it may originate. The term would be *nucleus idea*. As the cells in the human body, ideas seem to originate and function much as our nuclear cells . As cells transform and integrate with others into life-support systems so do ideas and images join together into meaning.

Another qualification is required. Thus far in our discussion, we have seen how firm is the Occidental faith in reason which has dominated thinking for much of the past two millennia. Our assumption that rationality is sacrosanct in examining concepts may be based on a halftruth. The inviolability of Western reasoning is called into question when we study prerational societies as found in MesoAmerica or traditional African cultures. Rarely had they ever conceived existence to be rational or the cosmos to be a logical system. Their forms of understanding usually used figurative language, simile, metaphor and proverb to express their beliefs about existence. These imaginative uses would hardly be called reasoning.

On the other hand, prerational societies are fully cognizant of natural phenomena. They are aware of metamorphosis and symbiotic relationships among insects, animals, and human beings, even though they may not use scientific terms to describe them. Societies imbedded in raw nature must have had an instinctual intuition of the invisible powers in nature as morphogenesis and symbiosis. It should be obvious that early man perceived his environment and conceived his world in terms of nature's bioprocesses. For him, to be in synchrony with them meant survival.

This anthropological excursion into prerational societies on different continents cautions us against our own possible, cultural myopia. When any society limits itself to studying solely its own intellectual roots, such as the Europeans' historical emphasis on rationalism, this singleminded perspective can severely limit the scope and universality of its knowledge. When any culture respects itself to the exclusion of those adjoining it, that culture is purblind to the intellectual and spiritual contribution of other societies to humanity's universal intelligence and moral conscience.

Metaphor as Metaphysical Meaning

If anything characterizes the beliefs, myths, religions, and wisdom of prerational societies, it is their focus on the figurative nature of language. In other words, prerational societies relied heavily on metaphor to discover the truths of life. Metaphors expressed their invisible bond to nature. Moreover, the symbols and archetypes that evolved out of their earthy metaphors became the focal point of their instincts, feelings and ways of thinking. Throughout traditional cultures of the world, the appearance of poetry, story and oral literature established metaphor as the central means of explanation. At times, metaphors joined humankind and the universe into one vast, mystical entity. Thus it is advisable to examine metaphor for its salient characteristics.

The title of Lovejoy's book is a metaphor because it suggests the thesis: "the entire universe is a 'chain of being.'" Since he uses this image to concentrate the meaning of his lectures, we need to understand what a metaphor is. To begin, "a metaphor is a means of analogy. Between things thought to be distinct, one discovers similarities in quality or essence."[3] (p. 118)

Metaphor is complex in that it "is primarily a mode of intensification and only secondarily of similitude." In addition, "Language used for its positive ambiguity implies widening spheres of significance." Indeed, "...metaphor is metaphor by reason of the multiple expectations and inferences it evokes." (p. 120)

Metaphor may function as mood, state of mind, or truth. When it evokes a state of mind, it "reflects a basic view of self or of the world." (p. 130) As observed above," a metaphor may also divulge a truth about life. The nucleus of a metaphor may be said to be a living truth, which not only perpetuates its multiple semantic associations but also assures the integral meaning of their implications." (p. 133)

Since metaphor can translate mood, state of mind, and truth about life, these appear to be stages of poetic visualization and show how the mind instinctively grasps experience. These are stages of discernment and degrees of significance. (p. 138)

The actual purpose of metaphor is to draw our attention to the interrelatedness of things. Often a poem uses the perception of one sense in terms of another. This is called synesthesia. "In the poet's imagination, the mysterious link of our senses invites us to discover more than what appears to be. The simplest sounds, smells and feelings make us aware of infinite things that restore our souls." (p. 141)

For instance, in Wordsworth's "Intimations of Immortality," the nuclear metaphor inquires into the relationship between human being, nature, and God. "The poet retraces our presentiment of immortality through man's natural life." (pp. 142–3) To Wordsworth, "imagination revealed the affinity of all things as well as the interdependence and interpenetration of substance and spirit. Through his poetry, he sought the universal laws that animate our being." (p. 143)

In John Keats' "Keen, Fitful Gusts" (1816), metaphor functions to express man's common bond: in honest toil, in self-respect, in independent thought, in good sense, and in worldwide brotherhood. (p. 143) Often poets perceive the inner reality of our common humanity.

Personification

Generally speaking, personification is an analogy made between a person and a creature or a force of nature. Often it is a projection of human feelings. Often nature is shown to empathize with human suffering and tragedy. Obviously the poet lives in an intimate relationship with nature far beyond what the common, modern man experiences. Personification not only demonstrates the poet's feelings. (p. 148) It serves to illustrate his sympathetic identification with the invisible powers in nature.

Symbol

In literature, a symbol is a sign of an unforgettable event. A metaphor changes into symbol when the metaphor acquires a history of its own, so that ambiguities are sloughed off and its unified, durable and complex significance has emerged.[4]

Prehistoric man's belief in the hidden unity of all forms of life may be the origin of metaphor and explain its ultimate uses: to ascertain and reaffirm that secluded unity.

Northrop Frye maintains that the world of mythical imagery is apocalyptic, that is, it is "...a world of total metaphor in which everything is potentially identical with everything else, as though it were all inside a single infinite body."[5]

Thus in meaning a metaphor can widen ad infinitum. Metaphor tends to the subjective when it evokes empathy for a hero or his experiences, but tends to the objective when it assists us to visualize the truth or the pervasive plan of life. It is in this latter sense that some of the best minds of the last two thousand years have tried to describe the mysterious powers and processes evident in the universe. They envisioned it as one endless chain of being.[6]

Friedrich Schiller

In the late eighteenth century, the great poet-dramatist, Friedrich Schiller, offered his own original worldview, thus transcending in good part the history of ideas before him. This German genius returned first to Plato's two conceptions of God. The first identified the Deity by His immutable, self-contained Perfection. The second apprehended the Creative Urge as realizing limitless possibilities through endless time.

Schiller gave the term *Formtrieb* to the guiding drive in nature which developed designs and perfected form. In other words, *Formtrieb* was nature's drive toward the total integration of form and design in its creations. Put succinctly, it described Being.

On the other hand, Schiller named *Stofftrieb* as the ceaseless drive in nature to create beings. Put another way, *Stofftrieb* was nature's amorphous, irrepressible urge to generate life in endless proliferation. Put concisely, Schiller sensed that the essence of Nature and of man were one and the same.

However, these opposing forces and processes seemed perpetually at war. The creative urge in nature seemed always in conflict with nature's need to control and discipline it so as to establish enduring designs and to perfect the forms it created. Schiller's conceptual impasse was probably due to the dualism inherent in the history of Christianity. This dualism had been designated by the term *psychomachia*. It describes the Christian's mortal combat in life between virtues and vices, freedom and guilt. For nearly eighteen centuries this theme preoccupied the European mind in religious art, literature and teachings.

Schiller himself created great works of poetry and drama. Through this creativity and apparently through his conversations and friendship with Germany's other great genius of the time, Johann Wolfgang von Goethe, Schiller apparently came to realize there was in nature and humanity a third power. This was the play impulse, *Spieltrieb*, which revealed a higher destiny possible in existence. It seemed a gift from God or Nature that men too often failed to realize in their lives.

If we look closer, Schiller's conceptions have marked similarities to the thesis of *The Creative Conscience*. His *Stofftrieb*, the drive in nature to create beings, is kindred to the morphogenetic force in nature discovered by microbiology. It is this process which produces the range and multiplicity of the earth's life-forms as well as the variety of knowledge and wisdom expressed in the sciences and humanities. It is the matter, substance, vitality and processes inherent in all life, art and thought.

On the other hand, Schiller's *Formtrieb* is kin to the symbiotic function in nature discovered in both microbiology and ecology. It is the restraining process in nature which disciplines life's creative vitality. Exemplified in humankind's conscience, our will power devises viable norms of expression and perfects mental and moral works. Thus in concert with creativity, conscience accounts for the sophistication of design and form in art, literature and philosophy.

The parallel between Schiller's definition of these two archetypal processes to the bioprocesses delineated in this book becomes clearer with further comparison. He describes "the selfcontained perfection of God," who accounts for all consummate natural designs. Similarly, symbiosis acts to complete the systems of survival of all living creatures. Symbiosis is also responsible for the evolved design of every living plant and animal. Applied to the evolution of humankind, symbiosis may be seen as responsible for our developed rationality and our symbiotically evolved conscience.

On the other hand, Schiller interprets the Creative Urge in nature as seeking the limitless realization of possibilities in time. In other words, impulse to innovation, experimentation and creativity manifests the natural process of morphogenesis, which *The Creative Conscience* will discuss, describe and define at considerable length.

Schiller's reinterpretation of the basic ideas inherent in the history of ideas, as elucidated by Professor Lovejoy, would be enough to consider the German poet and dramatist a philosopher in his own right. However, it is his transcending the dualism of the past which shows his own true, philosophical genius.

Schiller conceived of life and art as the marriage of impulsive creativity and prudent conscience, the free abandonment we experience in play. *Spieltrieb*, the play impulse is inborn in humanity.

This marvelous insight into nature and humanity evokes a number of tantalizing conjectures.

1. The urge to play may be understood as a dialectic between vital processes not only in human nature but in Nature itself.

2. In humankind, the play urge is expressed in the creation of masterpieces of art, music, writing and thought. It is expressed universally in ideas, emotions, passions and love.

3. On television, nature programs explain the play of young animals as practice for dominance and survival. This dour Darwinian emphasis misses

the basic need to express their frisky animal energy, the sense of joy at being young and alive. In the old evolutionary theory, there is no place for play and creativity.

4. Yet as applied to natural processes, the concept of playimpulse calls forth a quite startling, deeper insight. Is it possible that the instinct for play may actually reveal a microbiological truth? Over eons of time, is it possible that the urge to play may even have served as a creative catalyst to bring forth ever new cellular and genetic levels of life?

5. If so, though poets, dramatists and story tellers obeyed the playimpulse in all ages, it seems to have been neglected by the religious leaders, philosophers and theologians.

6. Indeed, it is probable that organic transformation and species' transmutation were the result of the seminal character of morphogenesis and the responsive nature of symbiosis. Together they brought life into being. Thus, the playimpulse periodically may have actuated evolution across the ages. Once necessity had taught life-forms the ways and means to survive, the experience of creativity may have become not only the greatest joy of being alive but also a new purpose for living.

7. If this speculation were to prove viable and valid, then another theory of evolution would be needed to supplement Darwin's contribution to our knowledge of life on earth.

A Biogenetic Basis for Philosophy?

In juxtaposing various historical definitions of an idea, we find succeeding interpretations to be both a disappointment and a promise of further discovery. Attempts at describing the enigma of the universe are bound to fail and succeed in part because, though incomplete, yet they search for universals that might bring us closer to some ultimate truth.

If we hope to avoid limitations yet to seek out larger meanings untarnished by time, we need to base our philosophy on scientifically verifiable facts supported by a logic that corresponds to the everlasting processes of living nature. Such is the hope of *The Creative Conscience*.

Unfortunately, modern knowledge is replete with "ologies," representing distinct forms of logic in conflict with one another or largely indifferent to what others truly mean. In addition, academe and science are saturated with disagreements and conflicting theories. At times there emerges an uneasiness that the ground of being of modern knowledge may rest on quicksand.

This insecurity seems due in part to the proliferation of scientific "isms" and a burgeoning of theories and formulas of indefinite value. Even the tried-and-true paradigm of reason, cum logic, no longer seems to reassure us of its inviolable validity. All too often, one proponent's theory is met with a thoroughgoing skepticism as to its worth. It seems near impossible to establish a universal truth, which the many specialized branches of knowledge can agree upon as a shibboleth through enemy territory.

Yet twentieth century biology provides an enlargement of scientific vision able to become a solid foundation for knowledge derived directly from animate nature. Interestingly, this neoknowledge emerged from a world hitherto largely invisible to us.

Of course, the reference is to ecology and microbiology. Until the final decades of the last century, we were essentially blind to the scope of the earth's biosphere. In that sense, we were blind to the infinitely great. On the other hand, the last century's advances in microbiology showed us what had for millennia been invisible to the naked eye. Hence we were also blind to the infinitely small.

This purblindness to the extent of our biotic environment should give us pause to reconsider many of our most self-assured assumptions about reality. Indeed, the discovery of two universal processes, morphogenesis and symbiosis, basically nonexistent to the uneducated eye, should make all of us stop and think about what we thought we knew.

Yet these bioprocesses are as omnipresent as the sounds broadcast by radio and the pictures emitted by television. Indeed, these pulsations of biological energy have penetrated us ever since the beginning of our hominid existence on earth. Like cosmic rays, morphogenesis and symbiosis permeate and transfuse us as waves of life. Only now we have identified the purpose of these invisible phenomena. We are just beginning to understand their biological significance to our being alive and their influence on our human destiny.

The two processes symbiosis and morphogenesis act in concert through all forms of life as the history of ideas had sensed intuitively. It appears that Being and Becoming interact dialectically to energize all living nature and each individual life. It is probable that they are the intrinsic source of evolution on earth.

The thesis of *The Creative Conscience* maintains that two bioprocesses represent the primeval driving forces throughout nature and in the evolved nature of humankind. By closely examining the effects of these bioactivities, we seem to gain a germinal insight into the processes of human evolution.

Futhermore, by studying their characteristics, we may gain an intuition into the reason for being of human life itself. It is hoped the following chapters augment understanding of nature's dynamics and intensify our focus of life's biogenetic meaning.

In *The Great Chain of Being,* Lovejoy defines the purpose of the history of ideas to be the investigation of beliefs, prejudices, tastes and aspirations of educated generations in the past. By extension, the intent of *The Creative Conscience* is to investigate how humankind's mind functions according to the archetypal processes inherited from nature.

It should already be evident that the mental processes associated with morphogenesis and symbiosis surpass in scope and intent the rigorous use of empiricism, rationalism and causal analysis. As valuable as these forms of reasoning have been in the past, they may have unnecessarily circumscribed the mind's greater capacities. Our ability to conceive the metaphoric meaning of the universe we live in also enables us to define the significance of human destiny.

A final comment on Lovejoy is necessary. In summarizing the import of his Harvard University lectures, Lovejoy concluded that the metaphor "great chain of being" had failed as a unit-idea. The reason was that it neglected the inner reality of biogenetic nature, that of Becoming. Moreover, the ratiocination used to explain the concept Being failed because the living world is neither static nor logical. Rather, nature is dialectical.

In the end, Lovejoy argued that the metaphor describing the universe as a chain of being was a failure. Nevertheless, his demonstration of the history of the idea was not a failure. Rather, his work is an admirable, highly sophisticated exposition of the limits of logic to illustrate the coherence, consistency and rationality of existence. The best minds in history failed in their endeavor because a metaphor is a different order of truth than inductive or deductive logic. Trying to argue that biogenetic life is rational is like building a majestic and elaborate sand castle that the next tide of scientific truth will wash away.

Yet Lovejoy also made clear the best thinkers in history sensed intuitively that life is a combination of Being and Becoming. If my thesis proves sound, it will show nature to be a dialectic of symbiosis and morphogenesis, a tao of Being and Becoming.

PART III

LIFE AS PURPOSE

3

NATURE'S POLARITIES AND WHAT THEY TEACH US

Darwin suggested that species diverged from a common ancestor. He defined the process as follows. "Natural selections lead to divergence of character;...the more diversified the descendants become, the better will be their chance of success in the battle for life." (p. 63)

He illustrated divergence from a prototype by noting the similarity in the bone structure between humans, cats, porpoises, and bats. He specified the likeness between our digits and wrist, the radius, ulna and humerus. Based on their resemblance in structure, they are declared homologous. Widespread throughout nature, it is assumed that divergence helped generate the evolution of species.

It is also likely that divergence in nature accounts for at least three manifest evolutionary phenomena. As differentiation, it led to speciation, the process by which distinct species are formed. In addition, it accounts in large part for processes identified as the specialization of organs and other physiological changes that enhance the potential for survival. Most remarkably, divergence probably led to the creation of prototypal body forms such as the cylindrical shape most evident in nature, radiata, mollusca, articulata, and vertebrata.

One of the most authoritative explications of Darwin's theory is that of Professor R.J. Richards in *The Meaning of Evolution*. (1991) His observation on Darwin's concept of the archetype is particularly worth noting. "The generalized vertebrate archetype was a primitive organism whose descendants have become specialized and differentiated through evolutionary adaptations."

Moreover, those descendants would thus constitute the myriad of families, genera, and species united within the vertebrate archetype." (p. 105)

The comment leads us to imagine how an archetype might generate so many different descendants. Perhaps it did so by originating a set of creative strategies capable of handling the expected and the unexpected in the most various environments. Moreover, the archetype, through its descendants, may accumulate a kind of memory. It would keep a sort of genetic record of its successes and failures, eventually teaching itself to meet the varied contingencies of survival.

If we take our speculation back further in time, we may imagine the condition of the earth at genesis. In the beginning, all was formless—a mass of matter, energies, and opposite attractions until a primitive polarity emerged when the continents pulled apart into a greater equilibrium of land masses in both hemispheres. Gradually, as atoms are want to do, the forces of nature interconnected. Out of the constant churning of the oceans through their gigantic underwater cyclones, life was stirred into being.

Hence out of formlessness, the urge to form came forth due to the polarity which magnetized every atom and bit of matter. That magnetism, created perhaps by the world whirling like a spinning turbine, created the electricity of life. Such electricity itself was polar, shaping the heads and tails of the tiniest form of life.

Yet sometimes, they took the form of radiata, articulata, and vertebrata. Although man and woman were latecomers in evolution, the relationship between semen and ovum serves as a useful model of the fundamental polarity. Those seminal heads with tails had to swim somewhere into life. Indeed, both semen and ovum seem to possess some kind of bioelectrical magnetism which draws them together to create new life.

What does this curious speculation lead to? Perhaps in the very beginning, no form of life appeared until the first amoeba and protozoa. From then on, however, the process of morphogenesis within each of the more developed forms of life became more articulate in terms of motility and mobility. This process expressed itself through divergence.

However, already amoeba and protozoa were the outcome of the integrating process symbiosis, which in cooperation with morphogenesis went on to create more viable, multicellular forms of life. Over a half billion years, there may then have developed an archetype or various prototypes similar to Wainwright's proposal that most forms of nature emerged as cylindrical shapes with heads and tails. From this elementary model, other life-forms eventually evolved based on the medium it lived in.

Corollary to the process of divergence was its convergence. Darwin was the first to use the term. Darwin suggested that species without a common ancestor may look alike because they had adopted similar environments, a phenomenon he called convergence.

Examples are porpoise, shark and fish. They all adapted to life in water and look alike, yet all evolved separately. Their fins, tails, and body shape he called analogous structures. Species with analogous structures are said to have converged. However, as Darwin pointed out, upon closer inspection such structures display obvious differences whereas homologous structures agree in detail.

Hence as a natural process, convergence means that distinct species have similar contours, which help the creatures survive in a common environment. It seems that the form enables them to conserve their energy; their concentric configuration gives them greater efficiency in moving through the medium. Indeed, the function of convergence in nature may be to consolidate and concentrate life's energies.

Interestingly enough, convergence may manifest a mode of intelligence characteristic of nature as it generalizes and universalizes life-forms. Moreover, distinct species probably evolved similar survival strategies, such as adaptation, predation, and evasion. Hence there must also have developed in each species corresponding patterns of perception and consciousness.

This was probably the case with primates and man. Homo erectus, Neanderthal man, CroMagnon, and Homo sapiens are obviously the same basic species. Because they faced common survival challenges, they must have converged mentally. Through evolving similar survival strategies, problem solving tactics, and modes of mental concentration, they practiced a new form of natural convergence. Through hundreds of generations of man, this mode of natural intelligence became humankind's ability to form generalizations and univeralizations based on their common experiences.

Divergence was probably responsible for the multiplication of multicellular organisms which were originated by the morphogenetic differentiation of cells. On the other hand, convergence may also have been responsible for the multicellular organization that produced analogous and homologous structures. Convergence would then be part of the symbiotic molecular cellular organic integration in all life-forms. This would account, in part, for characteristic mental patterns in humankind, divergence and differentiation as well as convergence and integration. Eventually such patterns of growth and consolidation would much later have an effect on humankind's capacity to evolve knowledge from empirical experience.

Divergence and Convergence in Nature

There are a number of inferences we may now make. At the purely biological level of understanding, multicellular divergence and convergence account for identical and similar archetypal structures and shapes among diverse species. At the same time, a corollary convergence of multicellular tissue and organs into life-support systems evidently helps guarantee the better adaptation of an organism to its environment. Yet it guarantees a corresponding homeostatic adaptation of the inner environment in every life-form.

Put in more general terms, divergence led to the diversification of a species' descendants. The infinitive *to diversify* means to give variety to. In Darwin's sense, diversification is a positive effort of nature to strengthen or extend a species' chances of survival. Thus it may be said that diversification is due to a lifeform's inborn curiosity, its instinct for exploration and experimentation, and possibly an aptitude for creativity. All these traits are characteristics of intelligence in nature.

Successive life-forms survived by integrating cells into multicellular organisms, thus illustrating the principle of convergence. By this principle the successful species learned to survive not only by avoiding gross and deadly errors but also by increasing their efficiency in survival strategies.

Convergence of cells into complementary tissues, organs and life-support systems helped to secure greater strength and flexibility in meeting life's threats and opportunities as it ensured a superior capacity of survival. On the other hand, the ingenuity and inventiveness obvious in nature, as energy, design and plan, enabled life to "hedge its bets" by creating and perfecting a wide range and variety of life-forms through the superordination of divergence and convergence combined.

For a moment, let us look more closely at the intrinsic evolution of life-forms and the processes which may have actuated that evolution. Divergence seems to lead to organic specialization and the development of a wide range of survival strategies. On the other hand, convergence seems to lead to generalization. This results, not only in cellular and somatic homeostasis. It also effects the enduring relationships among the organism's survival systems by getting them to function together to keep the life-form alive.

Divergence and Convergence in Human Nature

If we consider the consequences that such facts of nature have on human nature, we may draw some interesting inferences.

The manifestation of divergence teaches us the necessity of creating new modes of self realization and self-defense. Similarly, convergence shows us the benefit of using our natural conscience to monitor our excesses as well as to complete and perfect whatever we undertake to do.

Put in down-to-earth terms, we need to learn to compensate for our ignorance and inexperience and to counterbalance our tendency to intemperance, incaution, and foolhardy risk. We need to bring about a sensible synthesis of our spirit for adventure (our instinct for divergence) and our common sense as to how far we should go (our instinct of convergence). Life decisions truly require us to exercise our creativity and conscience. We owe this understanding to the intelligence inherent in nature itself.

Moreover, there is value and validity to the polar explanation of our evolution as creatures. The conditions of survival are best understood as due to learning to synthesize the polar tendencies of our dialectical nature and personality.

Nature's dialectic also characterizes our psychic and noetic processes. Understanding ourselves, as such, should enable us to guide our lives with more natural intelligence.

The traditional dualistic view of human nature was that *psychomachia* ruled our souls; good versus evil, virtues versus vices, morality versus immorality, God versus Satan. By contrast, to understand our nature as being fundamentally dialectical means we see life as a learning process with healthful and harmful phases teaching us their truth and value. So it is important to recognize human destiny for what it is in the context of our true nature.

The vision of life as made up of polarities of influence and attraction is important because it enables us to grow mentally, emotionally, and spiritually through each stage of life. We must not waste it in senseless conflicts of self with self as dualism would have us do.

For the human being, the divergence manifest in nature's most fundamental processes has a lesson worth learning. In human terms, to diversify may simply mean to seek out a broader range of experiences for the excitement and knowledge the new brings with it.

On the other hand, by investing one's life in different places, people and styles of living, one diverges from the past to learn what the future may bring. For the human species, diversification is likely to help us acquire a greater abundance of pleasures, adventures, and gratifications.

Obviously, for the human race, diversification of experiences goes beyond any theory of the "survival of the fittest" or "struggle for survival." Rather, it expresses the joy of being alive and being in love with life. It expresses our curiosity and creativity.

Correlatively, convergence in human nature draws us to consolidate and integrate life's experiences into sensible understanding and a more mature conscience. Perhaps this focus also gives us foresight into our future. In any case, together, divergence and convergence seem to lead us to a natural wisdom how to live life so as to discover a meaningful destiny all one's own.

A few final remarks are in order. Throughout evolution, polarities appear to give expression to a dialectic of natural processes. One polarity evinces divergence, exploration and experimentation, which accounts for complexity and intricacy of invention as well as ingenuity of human deliberation. The other polarity emanates convergence, concentration, and simplification, which together produce adroit design, efficient performance, and effective accomplishment. In this way, human nature exhibits parallel polar characteristics of creativity and conscience.

What do these polarities teach us about human evolution? Well, we are not simply survivors of a relentless conflict of life-forms nor are we the result of blind, evolutionary forces. As a species, we survived because Nature's Infinite Intelligence was transmitted to us as a finite intelligence composed of creativity and conscience. By that polar intelligence, we have been given the means of mastering the competitive powers of nature that would otherwise have made us an extinct species.

Our polarities help define our values. Life is creativity. Life is decision. Life is conscience. We are born to perfect and complete human nature. We are born to improve the human condition, our own and that of humanity.

Growth of Complexity Versus Simplification

In evolutionary theory, there seems to have been an implicit conviction that nature has overwhelmingly proceeded from the simple to the complex. Both in the Linnean classification system of the human and in Robert Whittaker's (1969) five-kingdom system, this overall view of species development seems to prevail.

In Linnaeus, Kingdom refers to Animalia as multicellular organisms; Phyl describes Chordata; Subphylum delineates Vertebrata; Class characterizes Mammalia; Order explains Primates; Family illustrates Hominidae; Genus distinguishes Homo; and Species represents Homo sapiens. Evolution is portrayed as a process of ever increasing complexity, proceeding from multi-celled organisms through adnexa, adaptations, apparatuses, appendages ad infinitum till present day Species are identified. (p. 20)[1]

Similarly Whittaker's five-kingdom system describes: Monera, Protista, Fungi, Plantae, and Animalia in an ascending order of cellular organization, means of producing their own food, and of reproduction.[2]

To be sure, the evidence is convincing. Over vast periods of time, the evolution of life-forms and species did move in the direction of greater organic intricacy and structural complexity. Of course, the evolutionary process implies the development of tissues, organs, and physiological features according to an organisms's needs. In fact, increasing complexity led to the creation of entire life-support systems. In other words, it meant that each stage of evolution signaled that the survival capacity of organisms increased.

Yet is this picture complete? Endless growth in complexity would be akin to proliferation run amok. It would seem that Darwin himself expressed his reservations.

"I suspect—some cases of compensation—a more general principle, namely, that natural selection is continually trying to economize part of the organization." Or again, "I believe natural selection will tend in the long run to reduce any part of the organization...causing some other part to be largely developed in a corresponding degree. Conversely, natural selection may perfectly succeed as a necessary compensation the reduction of some adjoining part." (p. 71)[3]

These quotations seem to imply that a counter process, simplification, must at times occur. If so, we must then ask what role simplification would have in evolution.

While complexification may be the primary principle driving the embryonic and ontogenetic development of the individual and while it may also be in effect through stages of evolution, yet there must come stages of maturity or completion when the process of complex growth would reverse itself to pursue simplification. The reason for such a reversal would be that any excessive complication of organization would ultimately lead to a "short circuit" or "blackout" of communication systems. Hence simplification would meet the organism's survival needs through initiating a greater economy of means and through stricter self discipline. In other words., complexification would be tempered, controlled and guided by simplification for the purpose of consolidating and integrating functions more efficiently.

To recapitulate, divergence and differentiation appear to be processes that initiate distinctions resulting in speciation and specialization in species. However, these same processes can also account for needless complications that endanger the continued survival of the organism. To prevent this, convergence initiates consolidation by simplification so that the organism continue to thrive.

But there is more. The process of simplification effectively "streamlines" each evolutionary stage as class, family, genera, and species. The corollary process of generalization adds successive stages of evolution to adapt to all variety of environment. As incarnate in the archetype of each species, its original capacity for generalization is probably passed on via cell, soma and genes through each stage of maturation and evolution. In such wise, successive stages are actualized and completed.

Put another way, homeostatic stages succeed one another until the present day life-form. Hence each species is a congruity of divergence and convergence, complexity and simplicity, specialization and generalization.

All this has an effect on our mental processes. Whereas divergence has taught us to differentiate, distinguish and create intricate systems of knowledge, convergence, has taught us to consolidate and integrate what we know into generalizations of grandeur and beauty. Hence through observation of corollary patterns of evolution in nature, we can not only define natural phenomena with some accuracy but also discover laws of both nature and human nature.

Thus as man's knowledge, experience and understanding have grown in complexity, so too has his simplification of nature's complexities led him to comprehend the principles, methods, laws and wisdom of his own mind.

Dualism Versus Polarity

Since ancient times, humankind believed existence was made up of polarities. About 600 B.C. Zoroaster, the founder of Zoroasterian religion, taught that the universe was a cosmic struggle between the supreme god Ahura Mazda and the evil spirit Ahriman. Again in the third century A.D., Manes, founder of Manichaeism, explained existence as a struggle between darkness (the evil material world) and light (the realm of the spirit). In fact, dualism has been characteristic of a good number of religions, including Christianity (God versus Satan), Judaism (its Angel of Death at Passover), and Islam (the jinn of Muslim demonology who exercise supernatural powers). Historical dualism held the conviction that the universe was controlled by two preternatural powers, one good and one evil.

Philosophers have also been known for their dualism. Descartes saw the universe divided between the physical and the spiritual. Kant discriminated between phenomena and noumena. Then again the ancient Greeks were vexed by another kind of dualism the opposition between change and Being. Indeed, Lovejoy's lectures showed us that duality tantalized and tortured good minds for two millennia.

However, today we know that animate Nature has little to do with any unearthly interpretations of existence and life. In contrast to the conflict of principles inherent in historical dualism, nature's polarities are composed of interacting and interactualizing processes, such as divergence and convergence or morphogenesis and symbiosis. Rather than opposites, they represent not only our biological potentials and limitations but also nature's injunctions against excesses and extremes.

Moreover, polarity of perception and sensible invites unity of apperception. In other words, recognizing the polarities of life teaches us there is a balanced, unified reality inherent both in nature and in ourselves.

Polarity indicates there exists a medium, a balance, a harmony, a homeostasis, an ecological oneness to the earth itself. There is a oneness between the human being and the biosphere, and ultimately, a symbiotic oneness of humankind's body and mind.

Polarity also points out that danger exists at the extremes as, beyond the limits of rationality, common sense, good sense, temperance and *sóphrosyné*. Polarity places man's apperception and moral sense in the center of our psychic symmetry guiding, controlling, harmonizing, integrating the outermost limits of life with the innermost capacities of the body-mind. Polarity promises sanity, homesostasis, reasonableness, tolerance, compassion, and "knowing thyself."

Evolution as Polar Process

In as much as nature is polar, it manifests the positive and negative, the creative and destructive, the integrative and disintegrative. Since Nature manifests itself in polar tendencies and directions, it is probably true that it activates a polar process between extremes.

When biologists speak of evolution, they marvel at the remarkable progress of species and especially of hominids evolving into Homo sapiens. Microbiologists have given us an understanding of how life built itself from the single cell to multicellular complexity and multi-organ creatures with amazing survival capabilities, yet evolution might be described differently.

Is it not just possible that evolution actually is an endless trying out the limits of life's polar extremes? We have already seen how the tendency to extreme complexity activates a countermovement towards simplification in order to create a more efficient and effective organism. In the cycles of nature, how far can disintegration go until nature initiates a counter movement toward integration? Think of her cycles of energy and her food chains. Think of the plants eaten by animals, and when the animals die they become fertilizer for plants. Think of the relationship between predator and prey. Where prey die off, so must their predators. As prey repopulate an area, more predator young can be fed so their population grows once again. Hence when the extreme or limit has been reached, a counteraction sets in to reestablish a balance.

But what do the extremes of existence teach life forms in nature? Perhaps they serve some purpose. Perhaps they may impede or limit all together the evolution of certain species. Or it may be the polar limits of reality are simply the way things are. No mystery about it nor any reason for it. In that case, existence would be simply mindless and meaningless.

With some reflection, a better answer is forthcoming. Nature's polar extremes have made it necessary for every living thing to be tested. For what? For the scope and limits of creature intelligence and the scope and limits of its life.

If so, then what about human beings? Humankind are prone to excesses and extremes in passions, convictions, and rash actions. We are continually involved in disagreements, accusations, homicide, war. Humankind are known as ultraists, fanatics, extremists, revolutionaries, terrorists. Some anthropologists are convinced we have descended from killer apes.

Then again another part of humanity is pressed into extreme poverty, desperate living conditions, incurable diseases, even suicide. In the modern world, violence is our daily bread, and mass murder of different ethnic groups, races, and religions is a commonplace. So at times man seems both villain and victim. How far can we go in that direction?

Without resorting to sanctimonious resignation stating that "the ways of God are mysterious and unexplainable," there may be another explanation, which is not a feckless attempt at consolation. If there is a natural reason for the polarity in human beings, it would be to find out who or what we are.

Any honest examination of our private and social life, culture and civilization reveals our polar tendencies. Any study of religion, history, literature or philosophy reveals the polar propensities of the human mind. As to the evolution of humanity, the most important fact would seem to be that human intelligence is seeking to learn its own identity or purpose in existence. Such identification would involve discovering our strengths and weaknesses, perceiving our limitations, and identifying the dangers of all extremes by finding out what we are made of.

Our identity would include what is reasonable, moderate, sensible, and moral. Identity is reached when the individual achieves a (homeostatic) balance of character and intelligence, desires and emotions. It also means coming to terms with the foul, the toxic things in us, which we need to learn to excrete so that our lives remain healthy, clean and morally sound. So it seems that nature's polar extremes are there for the purpose of evolving our essential human nature.

This means we must learn to listen to the comon sense and sanity required of human life. It means living in control of our excesses and keeping self-discipline in balance with self-fulfillment. Of course, that means having a private life without forgetting those who need our help. When we have gained a measure of equipoise by balancing and synchronizing our ambitions, desires and passions, we are once more ready to travel, to search out the new, and to create a new life. We will then be prepared to become who we want to be, to realize a destiny worthy of our personality and character.

Put in a philosophical context, perhaps our so-called evolution is in actuality a dialectic between belief and skepticism, faith and cynicism, dogmatism and hereticism, fatalism and freedom. Each of these extremes teaches us a bit more about ourselves and our humanity.

As the French mathematician and philosopher Blaise Pascal (1623–1662) would ask, "What of the eternity before we were born and the eternity after we die?" Indeed, what do these time limits teach us about the brief eternity we have yet to live?

Two Visions of Human Nature

We hold two contradictory visions of human nature. In one we visualize ourselves as competitive, contentious, homicidal because existence itself is thought to be a struggle to survive. That vision seems to illustrate relentless species divergence.

In the second, we envision humankind as converging through the creative and conscientious use of intelligence, through teaching and learning from one another, by pursuing common needs and goals, through understanding our common destiny of mortality. Convergence leads ultimately to mutual tolerance, mutual respect, mutual help, a destiny of shared purpose for all humanity.

Human history shows us that divergence of languages, beliefs, customs, traditions, mores, religions and life philosophies lead to misunderstandings, disagreements, confrontations, conflicts, fratricide, and ultimately to wars of mutual annihilation.

By contrast, over thousands of years, human history has shown that humankind have actually converged through the sensible exercise of human intelligence and compassionate understanding. Indeed, human convergence is as important an influence on human destiny as all others. There is the possibility humanity will converge symbiotically through worldwide reciprocity of knowledge, mutual aid and wisdom. Humankind will ultimately understand that our fealty to the Creative Conscience in Nature and in our beings can and will evolve us as human beings worthy of our creative intelligence and the symbiotic, sympathetic understanding of our own kind.

This has been intentionally brief. It speaks in general terms of intelligence in humanity. A more comprehensive and inclusive discussion of this topic will be found in Chapter 8, The Polarity of Human Nature in Culture.

4

NATURE'S PROCESSES SERVE A MUTUAL PURPOSE

The purpose of this chapter is to define and describe nature's two universal bioprocesses: morphogenesis and symbiosis. While the following exploration of the two key terms is divided logically for the sake of greater clarity, in nature they are inextricably intertwined because they actuate and actualize each other at given periods of need in living organisms.

The mutual coordination of these two bioprocesses requires us to keep their interaction in the foreground of our understanding. Not only are these processes manifest everywhere in visible nature. They are also evinced universally in the organic world within living creatures and in the microbiological world of cells and genes. Thus any attempted explanation of morphogenesis and symbiosis necessitates providing evidence of their presence in visible external nature and in "invisible" internal nature. In other words, beyond and beneath the perception possible with the naked eye, we must plumb the depths and intricacies of our own body and brain to be able to grasp how completely these processes affect our lives.

To initiate the nonspecialist into the vocabulary used by the modern biologist, let us first define our key terms and then provide sufficient examples that all can understand.

The easiest way to understand morphogenesis is via the outward manifestation of metamorphosis. Every school child has learned of the remarkable transformations that occur when a caterpillar changes into a butterfly or when a tadpole becomes a frog. It is a startling and delightful display of nature's hidden powers.

From that childhood experience we learn that metamorphosis brings about a marked developmental change in the form of structure of an insect, amphibian, reptile or animal, once it has been hatched or born.

But what force brought about such a metamorphosis? Twentieth century microbiologists discovered an ubiquitous process in nature that they identified as morphogenesis. We all know that genesis means the origin or coming into being of something, and of course, from the Bible we know that Genesis meant God's Creation of the earth and all living things. This Biblical interpretation was a figurative way of explaining a fact of nature obvious to all mankind, and all races have given us their own Creation stories.

Moreover, the dictionary tells us that the prefix morph or morpho relates to form. In other words, morphogenesis is the manifestation of the power of nature to effect changes in natural form, i.e., the evident changes that one can see in the growth of plants and animals.

The dictionary defines morphogenesis as "the formation and differentiation of tissues and organs." The adjective morphogenetic "is concerned with the development of normal organic form," e.g. the growth of an embryo. We are aware that human life moves from infancy through childhood, adolescence, adulthood, middle to old age. That too is a kind of metamorphosis. Technically, it is called ontogenesis, which means the course of development of an individual organism which we detect by their morphological changes.

Now, the greatest change biologists have discovered is evolution, a term well known to all. Nineteenth and twentieth century scientists wrote a greal deal about evolution, and they provided us with an abundance of evidence for its manifestation in numerous treatises and books.

An excellent work is Victor B. Scheffer's *Spires, of Form: Glimpses of Evolution* (1985)[1] What follows are a number of examples he has provided as external evidence of metamorphosis and evolution, which illustrate the outer, visible signs of the creative power intrinsic to all life.

His examples focus first on animal evolution. He discusses invertebrates, fishes, amphibians, reptiles, birds and mammals. Under the diversity of life-forms, he speaks of the inventiveness of Nature over eons of time, which accounts for the great variety of the world's creatures as well as for the richness and complexity of ecosystems. Such invention is the very essence of morphogenesis.

Elsewhere, Scheffer points out how animal life is a "…continuous exercise in problem solving." Nature is capable of resolving difficulties and overcoming limitations related to basic survival. In fact, nature even invented novel reproductive strategies. This capacity for creativity is another salient characteristic of morphogenesis.

In addition, most animals have methods of defending themselves. One defense is to assume a passive posture as does the possum "playing dead" when a predator appears on the scene. A much more common defense is the use of camouflage which conceals a creature in similar surroundings. Protective coloration hides many species from those that would eat them.

In addition to these and other examples of morphogenetic creativity, Scheffer cites a multitude of instances of symbiosis throughout external nature.

Probably symbiosis is a term more familiar to the reader. The dictionary defines symbiosis in two ways: (1) "the living together in more or less intimate association or close union of two dissimilar organisms; (2) the intimate living together of two dissimilar organisms in a mutually beneficial relationship, especially mutualism.

Several biologists have written valuable books on symbiosis. Scheffer also provides us with some salient examples. For instance, he notes that when the environment changes, those least able to change are eliminated. (p .4) Hence symbiosis influences the evolution or species. Those creatures which survive gain the capacity to change according to the demands of the environment. Such symbiotic response is inherent in most living creatures.

Life is found present in every conceivable habitat on earth. Moreover, "Life is an animate capability of forming and dissolving unions." (p. 11) Thus, in essence, symbiosis shows the capacity of creatures not only "coalescing and uniting with others but also of breaking up," separating, dissolving such relationships. Put another way, what becomes outworn or useless is discarded so the new can be found and integrated.

Scheffer also points out that nature's diversity is needed to preserve its stability. "A diversified ecosystem has overlapping checks and balances which guarantee that the system as a whole is buffered against the impact of any particular change."(p. 17)

As an illustration of symbiosis, he points to the hundreds of fishes, shellfishes, and soft-bodied organisms which interact in marine, tropical reef communities. Obviously these are "adaptations for survival." (p. 25) Hence, despite species competition in nature, symbiosis maintains its own form of control by checks and balances between them.

The term symbiosis is used in a more universal sense when it points to whole communities of creatures that seem to evolve as one great organism. In other words, "all evolution is coevolution" until the entire biosphere of the earth is involved. (p. 28)

In addition, he observes "...think of an unbroken thread of life that originated forty million centuries ago." (p. 30) The observation makes us realize there is another universal, ageless implication to the term. The unbroken thread refers to the intrinsic symbiosis that enabled life-forms to survive over the ages. Of course, this thread is at the cellular level of life. The stringing together of cells in complex, intricate and comprehensive symbioses to produce life-forms is the key to the durability of life across eons.

That fact seems evidence of a megasymbiosis at work on earth. That megasymbiosis is evidently the medium by which creative morphogenesis not only overcame the apparent inertness of matter, but also, together with symbiosis, perpetuated all the life-forms ever created.

There are many other examples of symbiosis throughout nature. "Mixed species foraging groups, observed among African grazing animals—gazelle, hartbeast, impala, springbok, wildebeest—forage together in various combinations, sharing the food source as they travel." (p. 39) Here we have an illustration of symbiotic tolerance, of "live and let live."

Further evidence of symbiosis and coevolution is the fact that mixed species routinely forage together in mutual tolerance On land, many birds, ungulates (hooved animals), primates, and in the sea, cetaceans (whales, dolphins, porpoises, and fishes) group together peacefully. (p. 39) Here symbiotic foraging and feeding predominate their ecosystem.

On the other hand, a sign of symbiotic "self control" occurs when birds lay only a certain number of eggs. Of course, too few or too many eggs may well cause the female bloodline to disappear. (p. 99) So clutch size is significant as an example of symbiotic adaptation to meet the survival requirement of proportionate reproduction.

A similar but broader implication is the increase and decrease in wildlife populations. This is symbiotic reaction to "environmental determinants as weather, food, predation and disease."(p. 101)

Examples of external symbiosis are more well known than those which reveal an internal/external form of symbiosis. Perhaps the most obvious evidence of the intimate inner/outer symbiosis in nature is "adaptation through compromise." The design of every plant and animal is based on a mutual adjustment between its needs and the means of survival in its environment. Indeed, for this reason it appears that "every animal is a multipurpose creation." (pp. 20–21)

This remark leads us to the reason for this chapter. Its aim is to make clear how nature's processes serve a mutual purpose. Hence it is time to explore how every living creature is actually the outcome of an ageless affiliation between morphogenesis and symbiosis within it own body and throughout all animate nature.

Morphogenesis and Symbiosis

We know already that each creature must live within physical limits of life, adapting itself to its own econiche in so far as it can tolerate the conditions. This fact makes us realize that all living organisms must develop the capacity to transcend limits in so far as possible. Diversity itself is not solely evidence of cellular, organic and species' ingenuity (morphogenesis) and adaptability (symbiosis). The interaction of these two processes enables the creature to transcend the limitations of past and present. Moreover, the two make it possible to circumvent, circumscribe, bypass and transform self-limits to new possibilities, new modes of being and becoming.

Scheffer points out that the ordinary agencies of mutation and recombination are capable of producing an almost infinite number of speciality designs (through gene expression). This entelechy comes about by the cooperation and coordination of morphogenesis and symbiosis.

For instance, specially designed organs for gathering food enable biologists to identify diverse species, such as teeth among animals and bills among birds. Furthermore, other counterparts as tentacles, talons and grasping hands make clear the kinship of other species.

Animal migration is a well known phenomenon since annually they migrate to special regions across continents.

Birds are especially interesting in this respect. Apparently they respond to the rotation of the stars around the pole, orienting themselves accordingly. It is thought that two hormones are responsible for reversing polarity, following the season. (pp. 53–54)

Otherwise, in general, birds are sensitive to polarized and ultraviolet light, to low frequency sounds…to atmospheric pressure, and magnetic lines of force." (p. 55)

On the other hand, pigeons have a geomagnetic sense of direction shared by certain bacteria, flatworms, molluscs, insects, sharks, salamanders, birds, mice, and porpoises. (p. 54)

Such phenomena provide the more obvious clues to the inner-outer correspondence between our morphogenetic/symbiotic natures and the relationship within as without and without as within all living creatures.

Put another way, morphogenetic/symbiotic intimacy exists between living organisms and their living environment.

The earth's magnetism is alive with energy. We do not see it, but the animals and sea creatures feel it. For instance, they know when an earthquake is coming well before our geophysicists can predict it. The sun too is alive with power because its invisible attraction holds together a solar system. And yet this source of celestial light radiates some uncanny form of energy able to call matter to life. Moreover, its constancy eventually helps create the multiplicity of life-forms, patterns and rhythms we witness all around us.

Creature orientation ultimately depends on its morphogenetic/symbiotic response to its immediate environment as well as its sense of distant places. Such orientation depends on a kind of compass inherent in various species and in man himself. Yet the capacity to orient oneself physically also implies survival orientation, the capacity to reach out or move toward some goal near or far.

A few more examples of inner/outer symbiosis are in order because they lead us more directly to consider the mophogenesis/symbiosis intrinsic to most forms of life.

Zoologists and ethologists study the instinctive behavior among animals. Today scientists consider such behavior as "genetically programmed."

With considerable insight, Scheffer himself observes that instinct limits the behavior of the newly born and limits later in life the ability to learn." (p. 6)

Hence when the instincts dominate, they may delay or even arrest the development of reasoning. Therefore, if human beings rely too much on instincts, in contrast to the capacity to learn and think, such reliance may limit the capacity to survive.

Scheffer comments further, "Behavior is guided in the 'lowest' organisms by instinct alone, in the 'higher' ones by instinct intertwined with learning, and in the 'highest' ones by instinct, learning and reason combined." (p. 6)

This statement has important implications. Note how "instinct intertwined with learning" suggests the need for symbiotic consolidation of instinct and experience. Hence to achieve a higher potential for survival, symbiotic integration is a vital necessity. Furthermore, if the "highest" organisms are to be guided by a combination of instinct, learning, and reason, the individual must seek to effect a more comprehensive, holistic, and complete symbiosis of his life. This means that our survival as an evolved species calls for the morphogenetic/symbiotic integration of human instincts, knowledge and reason. Clearly the understanding requires a new definition not only of animal survival intelligence but also of Homo sapiens sapience.

In essence, that means we should cooperate with nature's two archetypal bioprocesses and articulate them in terms of human aptitude, talent and intelligence. From our morphogenetic nature we should satisfy our curiosity and intuition for creativity. From our symbiotic nature, we should develop further our sense of self-direction, determination and decision so as conscientiously to decide the outcome of our lives.

Let us now probe deeper into intrinsic manifestations of morphogenesis/ symbiosis within life forms.

As Scheffer reminds us, "With the aid of the electron microscope, photo spectometer, gas chromotograph, and synchroton radiation probe, biologists have begun to 'see' things Linnaeus could not have dreamed of." (p 14)

Here we speak of the hitherto "invisible." It was always there, but until development of microbiology, we did not realize that morphogenesis actuated visible metamorphosis. Nor did we understand that symbiosis actually existed within living bodies. Hitherto we thought symbiosis existed solely in the adaptation of creatures to many environments, and more recently, in the coevolution of species.

In the Third Millennium we understand much better the five-kingdom arrangement of life. If the most primitive bacteria, *Monera*, showed the capacity to rapidly self-transform, such elementary forms of life evince primeval nature's morphogenetic power. The next level of cellular evolution was the algae and amoebas, the *Protista*; though one-celled, they displayed remarkable self-experimentation, which initiated the next step in evolution. Here is evidence of morphogenetic activity integrated symbiotically by a nucleus.

In green plants, the *Plantae* are capable of synthesizing biochemical compounds. Thus as symbionts in their own right, they continue to manifest the same penchant to morphogenetic experimentation. On the other hand, molds, mushrooms, and fungi display dependency on other organisms, which shows a further capacity of symbiotic integration. Finally, Animalia, typical animals made up of complex organic compounds, are active, sentient, and able to move in any direction at will. Hence animals represent the highest form of morphogenetic/symbiotic cooperation and coordination.[2]

This layman's thumbnail sketch of the five kingdoms of life allows us to draw some tentative conclusions. Evidently, morphogenesis is the creative capability of forms of life to solve survival problems. On the other hand, symbiosis indicates the capacity of life forms to adjust, adapt, exchange, mutually interact, and integrate past evolutionary experience with present existence.

A distinct and striking clue to the essence of the microbiological realm of life is the "double helix" configuration of nucleic acids DNA and RNA, which appear uniform throughout the living world. (p. 13)

The nucleic acids are localized in the cell nuclei. There, as the molecular basis of heredity in many organisms, DNA is constructed of a double helix held together by two chains of alternate links, and the RNA is in control of the synthesis of proteins.[3] In terms of my theory, the double helix would seem to actualize the two universal bioprocesses discussed so far. The DNA helix manifests the characteristics of both morphogenesis and symbiosis.

At this point in our discussion, let us quote a few remarks from L. Thomas's *The Lives of a Cell: Notes from a Biology Watcher* (1975). The quotations focus on inner organic reality.

To begin with, he describes "how the two organelles (of living cells), the mitochondria and chloroplasts are, in a fundamental sense, the most important things on earth. Between them they produce oxygen and arrange for its use. In effect, they run the place." (p. 84)[4]

He adds, "...these same cells are in seagulls, whales, dunegrass, seaweed, hermit crabs—and inland on leaves of beeches in my backyard. Through them I am connected." (p. 86) In fact, "We share genes, enzymes, and organelles with all forms of life." (p. 122)

In as much as you and I are connected intricately with all forms of life, it must be evident that we all embody life's archetypal processes, namely, morphogenesis and symbiosis.

Indeed, Thomas established this fact in two further observations. He notes that "the DNA in our nuclei may be from a fusion of ancestral cells and the linking of ancestral organisms in symbiosis." (p. 13) This morphogenetic/symbiotic link connects ancestral life-forms to today's. Indeed that line passes through geological time since it began 4000 million years ago: the earliest fish; the earliest corals; the earliest land animals and insects; the earliest vascular plants (ferns and mosses) and amphibians; the earliest reptiles; the age of dinosaurs, birds and reptiles; the earliest flowering plants; the earliest large mammals, grasses and hominids; and finally the earliest humans.

In other words, the double helix seems to represent the morphogenetic/symbiotic process which evolved all life.

The helix incarnates life's processes. Its design synthesizes, synopsizes, and systematizes human understanding of the symmetry and asymmetry in all life-forms, their conformity and regularity versus their imbalance and irregularity.

The sign announces that peridocially a kind of genetic "symposium" takes place over evolutionary time. It epitomizes synaptic exchanges for a mutual purpose. The helix symbolizes Nature's morphogenetic/symbiotic intelligence.

Put another way, the helix represents life's infinite power of survival by intelligent self transformation and by prudent integration of cells into tissue, organs, and systems so as to be able to sustain life in whatever form it takes. Evidence of nature's creativity and wise design is manifest throughout the earth in the history of invertebrates, fishes, amphibians, reptiles, insects, mammals, birds, placental mammals, and mankind today. In terms that the scientific mind can accept, life may be the only eternal power humanity shall ever know.

From a somewhat different perspective, Thomas states, "If it is in the nature of living things to pool resources, to fuse when possible, we would have a new way of accounting for the progressive enrichment and complexity of form in living things." (p. 33) From my point of view, the morphogenetic/symbiotic explanation is a more meaningful interpretation of the way evolution actually came about.

The ubiquitous evidence that morphogenesis and symbiosis activate and actualize each other is a convincing argument that nature's processes serve a mutual purpose. Obviously that purpose is to create life-forms as perfect and complete as possible.

A Morphological View of Nature

Stephen A. Wainwright's *Axis and Circumference: The Cylindrical Shape of Plants and Animals* (1988) provides us with a morphologist's or a mechanical engineer's interpretation of nature's forms. He states his objective thus. "The purpose is a study of the functional morphology of plants and animals. Functional morphology is the study of form, structure and function. and their relationship to one another." (p. 4)[5]

Obviously, the study of the intersection of form, structure and function assumes that nature's processes have a common purpose.

Wainwrght focuses on how the bodies of multicellular plants and animals are cylindrical in shape. Indeed, their posture and movement are physically based on the cylindrical form and its mechanical properties. Moreover, "Function refers to the dynamic physiological processes of organisms...and their structures over time." (p. 4)

Hence morphogenesis and function are intimately interconnected. What may begin as purely morphogenetic activity eventually evolved into a physiological purpose and correlative functions that strengthen the organism. Thus organization implies symbiotic design and integration with the purpose of increasing an organism's efficiency. Therefore, morphogenetic self-realization and symbiotic self-perfection together serve an ultimate purpose, the improvement of prospects of survival.

At one point, Wainwright cautions "Some biologists say that the structure of an organism is a system of constraints." (p. 5) But we must ask why not a system of new possibilities? Why not a system for discovering the most efficient means to effect such designs?

Wainwright's answer is that "an organism acquires new functional capabilities with each new structural mutation…" (p. 5) He illustrates his argument thus. "What all organisms have in common is a cylindrical shape of the body, roots, branches, appendages and internal conduits." (p. 59)

He then goes on to identify features of the materials and structural units that can account for the shape of all fossils and extant multicellular forms. (p. 81) One such feature is intercellular adhesion, that is, the crosslinking of polymers, which either attract or interact with one another in some type of bond…" (p. 83) He states further, "This crosslinking produces bone, coral skeleton, mollusc shells, which are the stiffest materials in the living world. Indeed, these ingredients of stuctural materials are found in plants and animals, past and present." (p. 88) Thus it would seem that symbiosis is basically responsible for the structures of animate life whereas morphogenesis is basically for functions. Moreover, over eons of time new symbioses generate new morphogeneses and new functions.

He also explains the probable molecular source of the cylindrical form. Macromolecules both inside and outside cells in plants and animals, assume a preferred orientation. That is, the cells grow in a preferred axial direction. (cf. anistropy as a functional property which creates a cylinder body)

Such orientation may be the consequence of morphogenetic experimentation and symbiotic design. Evidently it is the most efficient and effective prototype to guarantee the survival of creatures in need of motility and mobility. In addition, such axial growth confers polarity to animals with a head and end to their body. This sleek cylinder shape reinforces the argument that complexity finally requires simplification to attain adeptness in survival. Cylindrical bodies are streamlined.

A few of Wainwright's other remarks relate more directly to the human body and consequently to human intelligence. For instance he states, "Harmony is the embodiment of congruity."(p. 7) To be sure, human beings are aware of the strength, health, beauty and harmony of the human body. We do not usually think of it as a congruity, and yet it has taken shape so as to warrant survival in nearly all the earth's climates and ecosystems so it must have evolved a special cellular, organic, somatic, and cerebral congruity of evolutionary purpose even in the few million years that we recognize our forebears as human. In fact, our species seems to have undergone a rather startling acceleration of evolution. Yet the change has been secondarily physical and primarily mental.

Such change could not come about only by developing our creativity in coping with survival problems. As importantly, it would have come about by conscientious effort to establish congruity between body and mind. Hence mature decision would control rash actions and passions. Thus conscience would oversee and ensure inner congruity. Mental coherence, creativity, and integration nurtured and educated the "soul" of Homo sapiens.

Wainwright offers his own "engineer's" way of looking at evolution. "Assume that new structural features appear sequentially as though arising by mutations. Each structural feature conferred on the newly evolved form has a functional attribute that its predecessor did not have." (p. 81)

If we apply this assertion to the emergence of Intelligence in Nature and intelligence in humankind it evokes a striking realization. The assumption leads us to ask what function Intelligence has, if any, above and beyond survival and reproduction? Put succinctly, in so far as human intelligence surpasses preoccupation with survival, it becomes a new Force in Nature and, in time, possibly in the Universe.

What Morphogenetic Nature Shows Us

At this point, it might be helpful to cite well known examples to illustrate nature's two omnipresent processes. Thus what follows in the next pages are several illustrations of plant, insect, and animal metamorphosis and symbiosis.[6]

Plant Life

Morphogenesis/metamorphosis manifest the creative capacity of plant life in sundry ways. For instance:

1. In different climates, plants maximize the benefits to be extracted from their environment. They respond to it by originating a wide variety of leave shapes, vascular stems and tree trunks as well as root systems that extend outward and in depth.

2. Plants invent all manner of ways to spread their seeds by creating airborne and waterborne seeds. Others offer all kinds of enticing fruit. On the ground the tree's fruit and nuts are eaten by animals or carried away for hoarding. Forest animals hide nuts and seeds to get through hard-to-survive seasons (e.g., squirrels). Hence some nuts germinate far from the parent tree. On the other hand, birds carry seeds in their digestive systems to drop them far afield.

3. Plants also create flowers to be pollinated by insects wasps, bees, butterflies and such birds as the hummingbird.

4. Plants even evolve defenses against the encroachment of rivals. They exude a poison to ward off the encroachers.

5. Plants even evolve defenses against devouring insects as ants and caterpillars.

6. In temperate climates, deciduous trees transform annually as evident when their leaves turn from the pristine green of springtime to the solid green of summer to the vibrant brown, red and gold of fall. In winter these trees are bare. This visible metamorphosis reveals their seasonal morphogenetic awakening, flourishing and quiescence.

7. Another example of such morphogenetic activity is the fact that seeds do not sprout at the first seasonal rain. They wait. They remain dormant until the right amount of rain falls, the soil has the right moisture, and there is enough warmth from sunlight. Thus seeds delay germination until "their time has come." Note: To be sure, a number of the above are also examples of symbiosis manifest in plants.

Other Forms of Life: Insects, Fish, Animals

1. The best known examples of metamorphosis are caterpillars which change into butterflies or moths, tadpoles into frogs, and eggs into reptiles or birds.

2. Some frogs or reptiles turn white to reflect the sun so as to better endure the heat.

3. All feathered and fur-covered creatures evolved camouflage to match their environment. Camouflage illustrates how coloration to match the surroundings provides protection for would-be prey, but ironically offers the predator a similar advantage. Obviously such metamorphosis is actuated by both morphogenesis from within and symbiotic response to the environment.

4. In the sea, to escape the notice of a predator, some fish can change color by matching the color of the water or the background.

5. In Africa, the lung fish has evolved primitive lungs and can move over land and bury itself in mud like the frog. In case of a long drought, it can go to

sleep (cf. aestivation), wrap itself in its own mucous, and can endure for years in this state.

6. Also, in Africa, after a fire, flocks of storks move in for a feast on roasted frog, mice, and insects.

7. The pigmy mouse uses creativity and problem solving when it stacks pebbles in front of its burrow to drink dew from them in the morning. The dew forms from the cold night and the warmth of the burrow.

8. The male sand grouse has belly feathers, which the female does not. He saturates these feathers and takes the water back to the chicks, which bury their bills into the male's feathers to drink.

9. Elephants dig down with their trunks to water level to drink water. Rhinos are also known to drink from the elephant-made wells.

Morphogenetic ingenuity is also characteristic of birds and insects. For instance:

10. Birds are known to use cunning and deceit to survive. Some exploit their unsuspecting neighbors with considerable skill. Birds such as honeyguides, widowbirds, cuckoos, and cowbirds know how to place their own eggs in the nests from which they have ejected the owners. Based as it obviously is on knowledge of the habits and characteristics of other birds, this is a form of morphogenetic/symbiotic opportunism.

On the other hand, other creatures manifest true morphogenetic creativity. For instance:

11. Weaver ants use their spinning ability to make cocoons to stitch together folded leaves to build structures and complexes.

12. Fungus ants use fungi as gardens to feed their families.

13. The ability some birds use to weave unique nests demonstrates a further form of morphogenetic intelligence by tying knots to secure materials.

In conclusion, we may state that this "bringing together of raw materials" indicates a form of morphogenetic intelligence based on symbiotic knowledge of the creature's environment. As we will see below, symbiotic instinct and intelligence are also found in a great many, evolved life-forms.

Symbiosis in External Nature:
Insects, Fish, Animals

Many examples are actually evidence of the interaction of morphogenesis and symbiosis with the symbiotic effect being the most noticeable. Some examples are:

1. Spiders, praying mantis, and all sorts of bug change color and shape to be concealed in foliage with the purpose of catching and devouring other insects. This phenomenon demonstrates morphogenetic self-transformation and symbiotic adaptation to other forms of life and their surroundings.

2. In the sea, the same ability is characteristic of numerous fish which can change their color so as to make themselves "invisible."

3. Usually female birds are a dull, unnoticeable color perfectly adapted to their background. Obviously this attenuated color enables her to better conceal the nest with her body and wings so as to both hide her eggs and herself.

By contrast, the flamboyant male colors not only serve the purpose of attracting a mate of the same species but also may serve to distract a predator away from the nest and his mate. Apparently, in nature the males are more expendable than the females, which guarantee species survival.

4. Many herbivores and carnivores have adapted their coloring to match the browned landscape of a dry, grassy environment.

5. There are also examples of double camouflage. Herbivores have a light underbelly, therefore casting a fainter shadow on the underside, while having a tan or dark back to blend with the environment and to better deceive birds of prey.

6. Many snakes have perfect camouflage color and patterned skin design over the length of their body so as better to hunt rodents and other small animals.

7. By contrast, some snakes have vibrant colors to warn off their potential predators. Some poisonous reptiles also have bright spots on their dark scaly bodies apparently for the same purpose.

8. In wintry climates, polar bears are perfectly camouflaged against the ice and snow. Smaller animals as rabbits change their summer coat to the white fur that enables them to "disappear" in a winter landscape.

9. In color and form, crustaceans (lobsters, shrimps, crabs, and barnacles) have adapted effectively to shore life and the tides.

10. In the ocean, the sea anenome with its poisonous tentacles hitches a ride on a passing hermit crab, which offers free transportation. An octopus cannot attack the crab with impunity, and so both anenome and hermit crab offer each other a mutually beneficial service.

11. Sometimes fear, based on hereditary fear of the prey in the presence of the predator, may be considered a form of symbiotic knowledge. For instance, when the blue shark's sensors detect the weak electric current emitted from the fearful heartbeat of victim, there seems present an instinctive knowledge based on relationship over many generations of predator and prey.

12. Another instance of "double camouflage" is the fact that, when in the deep blue of the sea sharks are seen from above, their backs are dark, but when seen from below against the glittering surface of the water, their underside is light. Still, if the surface is very bright, they will cast their own shadow.

In sum, the evidence of morphogenesis and symbiosis in the relationships of a great many plants, insects, birds, reptiles, land animals and sea creatures lead us to conclude that morphogenesis-metamorphosis are manifestations of ingenious, evolved survival strategies. On the other hand, there is an undeniable interaction among all plants, insects, reptiles, birds, animals and sea life, which not only necessitated establishing intricate relationships among them. Such symbiotic interactions nurtured the coevolution of much of life on earth.

Other Manifestations of Symbiosis

In addition to the above cited examples of symbiosis among the earth's creatures, there are other manifestations. For instance, migration of fish, birds, and insects may be viewed as symbiotic atunement of species to forces of nature as gravity, magnetic alignments, stellar light, the tilt of the earth, or the location of food half across the world.

More well known is the polarization of male and female in all phyla and animals. Clearly the sexual act is the expression of a deeply felt form of symbiotic oneness. Similarly, the bond between mother and offspring is based on the strongest symbiotic instinct we know.

By contrast, at the smallest scale of visible life in external nature, symbiotic orientation is shown when ants and bees respond to angles of sunlight and light polarization.

On the other hand, many creatures are social in habits and life style. Insects such as bees, wasps, ants and sawflies are known for their cooperation, sharing and coordination of efforts for the larger community. Wesson considers social cooperation to be an evolutionary dynamic development of environmental adaptation.

Moreover, Wesson considers a community of animals to be "in effect more intelligent and more capable."[7] (p.135)

Yet of special interest is the evidence among mammals of widespread adaptability to the environment. Such adaptivity is essential to species survival. Ultimately, it depends on an instinct which incorporated the morphogenetic experience of the environment and symbiotic evaluation of that experience. Put another way, wherever there is instinctive or conscious use of symbiotic intelligence, the potential for survival increases.

Symbiosis is most obviously manifest by knowledge of the environment when animals forage, hunt, hoard food, or migrate. Such activities are clear manifestations of symbiotic hindsight and foresight.

On the other hand, the use of symbiotic instinct probably provides another means of survival as when animals herd together to establish more effective group vigilance for mutual protection or when select birds are on guard to emit warning signals to the flock at the approach of a predator. Such altruistic acts may be instances of an instinct of self-sacrifice for the group. In sum, any activity serving the purpose of mutual survival would illustrate the use of symbiotic intelligence.

Before concluding our evidence for morphogenesis and symbiosis in external nature, we should make a few remarks about ecology. To quote Barron (p. 303) "Ecology studies the relationships between living species and their physical environment. The environment includes living or biotic factors and the nonliving abiotic factors."

While the abiotic environment consists of its physical and chemical conditions, including water, oxygen, soil and light, each species has to learn to live with these environmental conditons in order to survive. Moreover, each species must interact with other species. Wherever mutual interaction takes place, there is a degree of symbiosis whether it be mutualism, commensalism, or parasitism.

However, at the core of all morphological change in each species, there is the interaction of morphogenesis and symbiosis within which constitutes their own intrinsic being and becoming. So in essence, if the bioprocesses influence the coevolution of all species in the external environment, as inherent biotransactions, they also affect and effect the evolution of each individual's internal environment. Moreover, these generic processes are present in every econiche, extending outward, sphere by sphere of life, to include the entire biosphere. In essence the entire abiotic and biotic environment is integrated for one mutual purpose to sustain and evolve life on earth.

In the following section dealing with Internal Nature, we shall find how cells, tissues, organs and life-nurturing systems within life-forms are an integral part of the individual. The bioprocesses we have seen actuate the relationships and coevolution of species are the selfsame that maintain human life. They not only actualize the embryonic development of all mammals, including man, but also promote the ontogenetic growth and maturation of the human being.

Microscopic Nature

The innermost transformation evident within life-forms is that small molecules become large molecules, then turn into bacteria and cells, and finally into plants and animals, to paraphrase Wesson. (p. 26) Hence the most evident fact of internal nature is its capacity to weave together strands of life into exquisite patterns of woof and web. Such intertwining creates the integrity characteristic of the symbiosis in all forms of life.

Called into action by germinal morphogenesis, symbiosis helps coalesce, consolidate, and integrate single cells into multi-cellular alignments, and stage by stage combine them into entire life-forms. Such integration was needed to establish a kind of "chain of command and obedience."

As evidenced even in single-celled organisms, the apparent aim of the two bioprocesses working in tandem is to protect, nourish and propagate the life-form. Yet the key to successful multicellularity is its creativity, i.e., the morphogenetic exploration of its environment, its self-experimentation, innovation and invention.

Proof of this creativity is provided by Wesson. He states "Self-reproducing mitochondria carry on metabolic processes. An outer membrane controls the passage of materials in both directions." (p. 135) Such absorption and diffusion reminds us of plant osmosis. It may also describe the flow of morphogenesis and symbiosis as a pervasive process throughout cellular nature.

Wesson also observes, a cell is organized to a high degree of stability in its entirety yet with possibilities of change. It represents determination coupled with unpredictability. (Wesson, p. 141, paraphrased)

Rather than unpredictability, a better term might be creativity to meet the unexpected and the unpredictable. Symbiotic design or homeostasis assures the cell's stability or stasis, yet the creative potential is always there to meet emergencies. This creativeness derives from its morphogenetic nature.

Each cell, tissue or texture and each organ or system of inner life has both a morphogenetic pole and a symbiotic pole. Morphogenesis is ever ready to initiate or to resume growth. Symbiosis is ever prepared to introduce repairs, to heal, cure and make whole again any damaged element, organism or system.

A mystical association comes to mind. With its finite membrane and its inner homeostasis, a living cell seems kin to a Tibetan mandala which is a graphic symbol of the circular universe, cradling at the center a Deific power who awaits birth. It takes human intelligence to find the ultimate meaning of both.

Wesson has another observation worthy of comment. "In living processes, positive feedback multiplies molecular instructions until checked by negative feedback." (p. 155)

This statement may be further elucidated as follows. Symbiosis provides "positive" feedback to morphogenesis when it encourages continued experimentation, innovation and growth, but "negative" feedback when it introduces the need for restraint and caution. Over time, this feedback became the organism's consciousness. In humanity, this conscious feedback evolved into our natural conscience.

The Gene Connection

According to the dictionary a gene is "an element of the germ plasm having a specific function in inheritance that is determined by a specific sequence of purine and pyrimidime bases of DNA or sometimes in RNA and that serves to control the transmission of a hereditary character by specifying the structure of a particular protein or by controlling the function of other genetic material."[8] (*Tenth*, p. 484)

Obviously it is not a simple matter to discuss this topic with such clarity that the nonspecialist can grasp with ease its characteristics, its purpose, and its importance to understanding evolution. So the following will undertake to explain the gene in the context of my thesis.

From the evidence given by the dictionary and various renown biologists, I would suggest that genes combine the binary influences of morphogenesis and symbiosis. Let me remind the reader of the essence of the two universal bioprocesses we have so far discussed.

Morphogenesis explores a whole range of survival possibilities and prospects, which effect intrinsic cellular mutation and variation. Symbiosis not only combines distinct genes but also consolidates the "lessons learned" from the morphogenetic exploration of the outer environment and from its own auto-experimentation with its inner environment.

Generally speaking, Wesson emphasizes the symbiosis inherent in the gene and in all systems of a living body. Thus he states, "to generate a new species requires not merely new genes but a new genetic cohesion." (p. 148) In other words, at the genetic level of life, the evolution of organisms proceeds by morphogenetic experimentation and by symbiotic consolidation. Moreover, genetic variations of DNA apparently undergo processes of transposition, duplication, and alteration which seem initiated by morphogenesis, whereas sequencing in patterns of growth is most likely effected by symbiosis.

Wesson makes another statement which acknowledges the capacity of both morphogenesis and symbiosis. "The power of the organizing principle of the genome is also apparent" in "the ability of a single cell to make a body of trillions of cells organized in thousands of organs." (p. 219)

Two insights may be inferred from this statement. The genome can originate trillions of cells not only through the power of morphogenesis alone but also through the cooperation and coordination of symbiosis which organizes new cell combinations according to the principle of consolidation. Symbiosis thus designs near perfect and complete complexes of multicellular entities. Of course, these compose the blood, tissue, arteries, nerve and all the systems that work together to give birth to and maintain a living body. Hence in this single example, we find illustrated the near omnipotence of the processes of morphogenesis and symbiosis to shape infinite forms of life.

One must wonder if the genetic code was fixed and could not be changed, as Wesson asserts on page 57. If fixed, what made it perfect, once and for all? Given the experimental character of animate nature, such fixity seems unlikely.

Moreover, how could evolution have taken place, unless genes periodically achieved greater effectiveness in perfecting nature's morphogenetic capacity to meet the challenges of a changing environment as well as continue to guarantee continued life? Moreover, since symbiosis had periodically to design and redesign forms of life, it did so in response to the impetus of morphogenesis, which generated life in the first place.

Wesson insists that the outer environment does not have that much influence on the species. He states a species represents an effectively integrated genetic combination, or attractor, likely to persist with little reflection of environmental change. (p. 198) Hence a species is primarily a genetic combination which is a viable or successful symbiotic entity, or put another way, a gene is a cluster of energies, signals, and bioprocesses capable of further generation and integration.

Wesson also remarks that the neoDarwinians don't believe in any direct influence of the environment on the genes. (p. 122)

Thus equal emphasis on the morphogenetic/symbiotic processes active within an organism and on the organism's interaction with its corresponding morphogenetic/symbiotic environment may be worth noting.

Wesson also informs us that most proteins are made by multigene families—homologous genes—which a single gene may turn on or off. (p. 39) If this is the case, a gene may switch on the entire process of morphogenesis (as the growth of a child) or may switch off morphogenesis so that the next phase of symbiosis may be switched on and left on till its work is completed. Then morphogenesis can resume its task in the further refinement and sophistication of growth, as when a teenager blossoms into adulthood.

One can only imagine if there is a gene or set of genes to meet the problematical and the unexpected in life.

Indeed, as the genome itself, each individual creature demonstrates self-direction and sustained effort at survival. That alone should justify our discarding any notion that "accident" of heredity explains the multitudes of species. Can what is born "by accident" suddenly manifest such self-determination and self-direction as cells, tissue, organs and life maintaining systems of the earth's creatures?

Moreover, the philosophical argument that life is predominantly a dichotomy of inner and outer seems an obsolete distinction, a conceptual confection of our heritage of dualism.

Wesson himself unintentionally suggests the bridge across this antiquated understanding of nature and man in nature. He does so in speaking of cellular activity. "Not only do cells receive a set of instructions telling them how to build a new organism; throughout life, instructions change." (p. 228)

If such instructions change over life, they must be in response to the organism's growth *and* maturation and its correlative change vis-a-vis its environment. Hence subsequent morphogenetic instructions would initiate subsequent stages of symbiosis based on its greater sureness of self and greater knowledge of the world, which would encourage further growth, exploration and experimentation.

When one changes, the environment appears to change. As every adult knows, year by year the world becomes a different place because we ourselves mature and age.

Put another way, the intrinsic interaction of morphogenesis and symbiosis accounts for both individual ontogenesis and species evolution. Morphogenesis proceeds through phases of growth and stages of transformation until a new need for consolidation and integration occurs; and then symbiosis satisfies this need. Moreover, genes operate together symbiotically by coordinating and combining their activities for the common good. Evidently their mutual purpose is symbiotic survival and morphogenetic reproduction.

All this activity shows the advantage of the multi-cellular organism over the single, isolated cell. This enhancement may be illustrated by a naive example. Think about the formation of the human body. Think of the evolution of bone, muscle, arms and legs as well as our cylindrical body which allows us to be flexible, motile and mobile.

This mobility through additional appendages are mutations from the unicellular world where we began. The evolutionary innovations enhanced our ability to survive virtually anywhere to feed ourselves, to fight, to flee, to explore, to experiment, to reproduce our own kind. Our limbs alone give us enormous advantage over the single cell. In fact, in a way, they are the most striking evidence of our morphogenetic evolution and symbiotic adaptation to our present world.

But what about the human mind? Since genome intelligence throughout nature makes itself manifest through morphogenetic/symbiotic activity, this fact allows us to speak of the morphogenetic/symbiotic intelligence of mankind.

In a way the human mind is our master gene nurtured by who we are but concealed in a secret world all its own. True, at times, it helps us discover what mind is through dreams, self-discovery, creativity, autobiography, integration of our emotions, intelligence and conscience.

To close this section on the role of genes, a few last remarks are in order. It would seem that some genes are produced as clones to strengthen and reinforce established organic and physiological structures. Many other genes perform specialized tasks. A very few could be characterized as "genius genes." They seem to have freedom from practical responsibilities or organic functions. So they use their "time" to invent, innovate, and create, perhaps for the sheer pleasure of it. (In animals that have survived, such genes might account for their resourcefulness, tenacity, and ingenuity. In Homo sapiens, they might account for human cleverness, creativity, and conscience.)

In the next chapter we will be occupied with how nature's processes created human intelligence. For now, we know that our intelligence originated in our primate cells. But further back in time, we suspect that certain primeval cells initiated life under the guidance of the morphogenesis and symbiosis at the nucleus center of animate Nature itself.

Tentative Conclusions

Both morphogenesis and symbiosis play vital, distinctive roles in embryonic development and in species evolution. Yet each reciprocates the action of the other, and together they effect a mutual purpose.

Morphogenesis as Polarity

As we have learned, morphogenesis in an organism may be said to initiate cellular growth and to explore the animate environment within a cell, organ, or life-form. The organism itself explores the environment without, that is, the raw physical environment, the condition of life and death in the surroundings. Morphogenesis searches out the life-giving and evades the death dealing.

Morphogenesis accounts for the cellular organic somatic growth both during the development of an embryo and throughout the lifetime of an organism. Put another way, morphogenesis emanates life energies not only to actuate general physical growth but also to actualize generic, archetypal patterns that, over time, will produce subtle variations as well as major transformations. Probably, morphogenesis is the ultimate source of species evolution as transmitted through cells, genes, organs, bodily functions and evolved structures.

As one pole of life, morphogenesis interacts with its complementary opposite, symbiosis, to complete the cycle of generation and consommation that the two natural processes enact together.

Symbiosis as Polarity

We have already provided examples of the work and effect of symbiosis in the external environment and within the internal nature of forms of life.

Although there are spheres and stages of symbiosis throughout animate nature, the overriding interest here is the fact that mutualism pervades much of life. Briefly, the term mutualism designates the mutually beneficial association between different kinds of organism, between members of a single species, and, between different species. Direct or indirect, intentional or unintentional, mutualism fosters mutual survival.

Given the intimacy and immediacy of relationships among all life-forms, they must mutually influence each others' lives, such that there exists a coevolution among all creatures in the same ecoenvironment.

The abiotic environment also manifests continual cycles commonly identified as the water cycle, the carbon dioxide-oxygen cycle, the nitrogen cycle, etc. Of course, these are the source of plant life germination, growth, flowering, and reproduction, which in its turn influences the animal "food chain" and the entire web of animate life. These continual cycles must also influence symbiosis throughout nature. On the other hand, at the microscopic level of life, homeostasis is the quintessential example of symbiosis. Homeostasis is defined as the state of equilibrium between different elements in an organism. As an interacting integration of mitochondrion, vacuole, cytoplasm, nucleus and so forth within an enclosing wall, a live cell admirably illustrates the accomplishment of symbiosis. At the same time, consolidation of multicellular structures, organs and functional forms of life exhibits the fact that symbiosis pervades all life.

However, as we already know, neither morphogenesis nor symbiosis acts alone, but mutually depend on the energy and purpose of the other. Their reciprocity may take place in the following way.

Morphogenesis and Symbiosis as Dialectic

Within an organism, once morphogenetic activity grows quiescent, because its energy is temporarily spent or because symbiosis must periodically exert its control, only then does the symbiotic process take over. Symbiosis next initiates the coordination and consolidation of the cells, tissues, organs and somatic systems that morphogenesis has created generically by preparing cells into patterns of possibilities. In some way, symbiosis transforms these patterns into viable forms of living substance with viable functions.

Once such an arrangement has been made, symbiosis then relaxes its control, which relaxation stimulates the reactivation of the morphogenetic process. However, it should be noted that this stage of morphogenesis is no longer the same elemental energy as before. Rather, the morphogenetic process has become more capable and sophisticated by reason of the integration effected by symbiosis.

Furthermore, by reason of its accumulated symbiotic knowledge and sophistication, morphogenesis is now empowered to be more skillful, resourceful, and more adept at creativity. It is ready to fashion more adroitly organic patterns, functions and forms.

So stages of embryonic development may be the consequence of the polar interaction of these two distinct, natural processes, each with its own particular function to perform. The two work together. To put it figuratively, morphogenesis and symbiosis transfer the mysterious energy of life into the embryo, irradiating the mother with instinctive love. In its own time the infant will emerge to bear the light of human intelligence into the world.

The Mutual Purpose of Nature's Polarities

To ensure the survival of life, nature establishes effective checks and balances on all manifestations of growth and on all cellular, organic, and structural consolidation of form.

Nature's polarity both invigorates and controls natural processes. As noted in chapter 3, divergence and convergence interact cyclically to foster and regulate each other's proceedings. The original, archetypal unity of the single cell evolved into multi-cellularity and, consequently, encouraged the continual growth of complexity. Nevertheless, this growth of somatic and organic complexity was checked periodically by a reverse process that of simplification. The purpose of such symbiotic simplification was to guarantee the symbiotic integrity of the entire organism.

If morphogenesis has the power to effect ever more ingenious and intricate configurations, symbiosis, by contrast, has the responsibility to resist excesses, to shut down dysfunctional networks, and eject malignancies. Thus, it is the task of symbiosis to ensure the economy and efficiency of its design so that the evolved form, whatever species it be, would survive.

For a moment, let us pause to concentrate on one microscopic phenomenon in nature, the homeostasis of a live cell. Do morphogenesis and symbiosis serve a mutual purpose in establishing the cell's perfect inner balance and harmony? If so, what would homeostasis mean in human terms?

In general, in the individual cell, homeostasis means ensuring optimum attunement, congruity and health. Obviously, such equipoise of cellular elements and functions could only come about through the reciprocal stability effected by morphogenesis and symbiosis when together they had perfected a life-form.

In human terms, that would mean we should seek to harmonize our bioneeds, desires, drives, and purpose in life. Homeostasis would mean the coordination and subordination of our emotional and moral needs. To achieve a sound destiny of our own, we would need to define who we are and pursue some lifetime purpose. It would mean identifying and integrating our state of being and becoming.

Rather than homeostasis being regarded as a too great preoccupation with self, it can mean an unselfish concern for the welfare of others. By extending the concept outward to include family, country and the human race with whom we share this earth, we may begin to feel our homeostatic oneness with all other human beings.

Sometimes this purpose can be fulfilled by reproduction, which requires the parents to assume full responsibility for the healthful, creative and moral education of one's children. Sometimes it would require the sacrifice of more selfish wants to the greater good of those in need. Sometimes it means creating better conditions of life for all of one's kind. (This is the parental instinct at its best.) To gain a sense of harmony and balance in our lives, we need to think of how we may mutually live together in peace to serve a common purpose for the good of all.

In human terms, homeostasis can mean not only the coordination of one's strengths and abilities to ensure personal survival but also the subordination of one's "superiority" to controlled, reasonable conduct befitting a human being.

Human homeostasis also means learning from one's successes and failures to maintain a sensible perspective on one's achievements and setbacks. It teaches us no single failure is the end of life. Even a series of failures has a lesson about the means and meaning of success.

Human homeostasis also can mean exercising hindsight, based on coming to terms with the truth of memories; and it can mean cultivating foresight to anticipate future requirements, problems, and the consequences of our decisions.

Human homeostasis also means integrating one's ephemeral life energies to give meaning to one's earthly destiny. Human homeostasis means passing on one's common sense and wisdom to one's descendants.

For humankind, nature's processes do seem to serve a mutual purpose to lead us to pursue an emotion-fulfilling, intelligent life.

PART IV

INTELLIGENCE IN NATURE AND HUMANITY

5

INTELLIGENCE IN NATURE

Present day knowledge of animate nature makes us realize there is concealed and revealed intelligence in the living world. In our studies of the microbiological realm, we witness intelligence in microbes, bacteria, and cellular systems. Moreover, the myriads of creatures inhabiting land, sea and air show by their remarkable adaptations that they possess degrees and forms of intelligence. Furthermore, the more we know about plant life, the more we see their capacity for survival in their successful adaptation to virtually all the conditions and climates of the earth. That too must be acknowledged as a form of tenacity to live. How else do we account for the planet's multitude of life-forms, species, genera and phyla than ascribe it to some form of conscious intent in Nature?

For most laymen as myself, it may be hard to consider the possibility that amoeba, protozoa, bacteria and viruses show signs of intelligence. In fact, some biologists prefer to account for microscopic life-forms as capable only of reacting to each other biochemically. Indeed, some scientists seem more comfortable to describe human life in biochemical electrical terms, especially in response to our physical environment. Fortunately, there is some evidence that humankind are more than SciFi androids.

While anthropologists retrace our skull transformations from our simian forefathers to *Homo sapiens*, evolutionary biologists assure us our species experienced a remarkable acceleration in evolution over the past few million years. Then there are micro-biologists who would point out that the human body and brain are made up of countless (intelligent) cells organized into systems serving one main purpose: to sustain our lives and keep the brain alive. So, possibly, human intelligence is a fact of nature too.

Because scientists are committed to scientific explanations, they no longer can accept the presence of the Divine in nature as the source of any intelligence it may manifest. Yet they cannot deny the fact that everything that lives seems to display some sort of intelligent purpose. Furthermore, scientists would agree that plant and animal cells, living organisms and every kind of conceivable creature on the face of the earth manifests selforganization; the ability to learn; the development of survival skills and strategies; and the aptitude to perform at least one function aptly. Eventually all these activities prove to serve one ultimate purpose.

Thinking Learned from Nature

Thinking eventually discovers itself by pursuing a purpose. In as much as the design of every living creature has the purpose of staying alive so as to reproduce itself, all animals must ACT in order to survive: suckling and feeding the young, foraging, hunting, finding shelter, and the like.

Where there are seasonal changes, animals face choices: some migrate to a part of the earth where there is food aplenty. Others remain in the same geographical area to endure the dry season or cold winter as best they can, but frequently they die of exposure. Those remaining in familiar territory "learn" to hibernate in cold climates or aestivate in hot climates. Others obey their instinct to hoard food to get through long months of privation. Others live underground where plant roots serve as a long term source of food.

In all cases, such plans to survive another year developed rudimentary "mental" aptitudes. Animals which did not learn to respond to the "inner call" of survival became extinct as species. To be sure, predators presented dangers that taught the pursued to expect the unexpected.

Natural disasters or the constant threat of predation taught them hard lessons. Memory may have oriented them to what the future might hold. At least intelligent animals did not remain fixed in the present, its abundance or temporary safety. Instinct told them to prepare for the unpredictable. This form of intelligence may even have engendered thinking of alternate escape routes or plans to move on. In any event, intelligence led them to think beyond the satisfaction of hunger and thirst and the need to reproduce.

Animal intelligence must have developed different ways of solving survival problems. Indeed, animal behavior enabled humankind to infer methods of thinking that we adopted. (Folklore is replete with examples and lessons of animals with specific survival skills or characteristics, such as the fox, the snake, the lion, the eagle, the wolf, the rabbit, the tortoise.)

To be more specific, foraging animals became adept at searching out the edible and the safe whereas predators adopted stealth and cooperation to ensure a kill. On the other hand, as the hunted became more experienced in detecting their killers, the predators became more skilled in stalking and bringing down their prey.

Both prey and predator used their natural camouflage to conceal themselves and to make the most of their senses to detect the presence of the other. Similarly, humankind itself learned to practice camouflage detection so as to evade predators or to discover the prey they hunted. Hence the use of camouflage detection and the ability to discover the hidden must have nurtured an elementary form of reasoning. Since detection led to inferences about the experience, the process became known as our mental habit of induction.

But let us consider further what animal intelligence may mean. A case of instinctual thinking occurs whenever creatures burrow out a "home." Some birds peck out a hole in a tree for its nest, and prairie animals dig tunnels of underground lairs. These activities are ways of hiding from predators or taking shelter from severe elements of the climate. Of course, it usually is home for their young.

By analogy, humankind may have emulated the "common sense" of animals by using caves or by building rude shelters. In man, this pursuit of practical goals led to the development of characteristic mental habits. Hence, these parallel developments of mental behavior incipient in animal and conscious in man became methods of thinking for us. Habitual detection of the hitherto unknown became the inductive research and reasoning esteemed by the sciences.

On the other hand, food-hoarding animals taught us another form of common sense reasoning. In fact, these habits may represent a step further up the hierarchy of mental skills. Food storage requires foresight to meet harsh conditions of life, such as surviving severe drought or freezing winters. Of course, use of foresight implies planning ahead, which means, in turn, pursuing a survival purpose.

In terms of human activity and mental habits, foresight depends on hindsight so that they teach us to coordinate knowledge of past and present in order to secure the future for ourselves and for our children. Indeed, it may be that traditional societies value hindsight because it has provided the benefits of the present, whereas "advanced" societies value foresight because planning is an adventure into the unknown to see if our plans turn out well. Put another way, traditional societies lay more stress on the moral value of conscience. Modern societies emphasize creativity and ingenuity.

Modern societies continuously seek to create a healthier, better educated, more abundant lifestyle for the individual. In other words, modern man seeks to ensure a fear-free society where each and every individual has the opportunity to create his or her own destiny rather than be fated by the mistakes, misdeeds, and miseries of our forefathers.

Hence, in a very preliminary way, humankind have learned direct and indirect lessons from our animal kin. Perhaps over time, the survival strategies of animals taught us how to perceive reality, to gain control over our immediate surroundings, to foresee the needs of the future, and to develop the mental habits necessary to secure our individual destinies.

If so, what has nature taught us?

In order to be alive tomorrow, we need to learn to protect ourselves from those who would harm us. Thus we need shelter for ourselves and for our own. We need always to hone our survival skills.

Animals know there is a harvest season of fruit, nuts, seeds, and other food, followed by a season of dearth and death. Humans also need to recognize that the stages of life offer their own harvest seasons to reap their benefits, but these are inevitably followed by old age and the winter of life. Hence there are stages in life we need to learn to save so that, when we age, we have nourishment and wisdom to live out well the last decades of life.

So it is probable that some of our most basic forms of reasoning have been learned from kindred creatures of the earth. They have helped us understand how best to survive and how to expect the unexpected that life may bring. Patiently we learned to learn inductively. Slowly we learned to discover the meaning of our experiences.

Humankind also discovered how to use hindsight to test the soundness of foresight. Like the major premise in a syllogism, hindsight tests the minor premise (our new ideas) to see if they conform to the common sense conclusions of past experiences. This procedure provides us with a deductive form of reasoning. Yet deduction itself must always be tested anew by induction, i.e. in the light of new facts and experiences. Thus do hindsight and foresight interact, dialectically as it were, to arrive at more mature and sound conclusions about our past, present and future.

Mothers in Nature

Thus far we have neglected the most important "force" in Nature—the maternal. In the animal queendom, virtually all "mothers" have acquired sufficient skill to assure the survival of their young. Not only does this kind of "thinking" guide their maternal purpose but it also indicates the clear development of instinctual foresight.

Animal mothers not only prepare burrows, find food, and care for their offspring. More importantly, the mothers represent the "conscience" of their species in distinct ways. Foremost, they are prepared to defend their young against all predators. They teach the young the skills they will need to survive. Their infinite patience is counterbalanced by disciplining the inexperienced by swift punishment when necessary. The mothers expect obedience and will not tolerate disobedience, for it may cost the young their juvenile lives. When the mothers and their progeny move in herds, a dominant female with obvious survival wisdom keeps the group from danger and leads them to sources of water and food. These examples are just a few instances of animal love and mutual respect in any given species.

At least so it seems to the human observer. These facts impress us as a law of nature more universal than that of predation or any struggle for survival of the fittest.

Among humanity, mothers evidently share many characteristics with their animal counterparts. As with animal mothers, the human mother is directed by a powerful, instinctual conscience. Motivated by a sense of responsibility, human mothers not only protect their young but care about all children. In particular, the human mother feels a deep need to help the hungry, the helpless, unfortunate, handicapped, and the homeless.

Such maternal morality is what is meant, in part, by the term symbiotic conscience, which is motivated by sympathy and charity for all. In a sense the word humanity really means the humane love of all mothers. It is to them we owe the deeper understanding of symbiosis as the source of humankind's natural, moral, conscience.

Forms of Survival Intelligence

Some degree or form of intelligence is manifest by insects birds, animals, and man. Reasoning of some sort may be guided by instinct such as when a spider weaves a web, a bird builds a nest, a beaver constructs a dam, or man puts up a shelter. Creatures develop a range of other survival skills which include the use of wile, guile and cunning to hunt and outwit predators, to hide and preserve food, and the like.

How did protohumans survive?

Very early on they survived by scavenging, which meant searching an area for any available food, dead or alive. (Gradually, they discovered that roasting meat kept them from getting sick.) This explorative activity led to a useful kind of learning of the habits and habitats of other animals.

Men learned to hunt which required them not only to scout the terrain but to observe with great care the specific characteristics of animals. The search for the facts of nature and patterns of animal behavior may be the forerunner to the inductive observation used by modern biologists and zoologists.

Then came the planning stage of the hunt. (The cave drawings of prehistoric man show how men hunted bison, deer and other game. Such drawings were likely the depiction of plans for the hunt as well as the use of images to invoke luck, called by anthropologists "sympathetic magic.") It is probable that such planning with a purpose eventually evolved into the practice of thinking that philosophers today call inductive reasoning, because it too was based on careful observation and conclusion about the facts.

For early man, the plan preceded the hunt. Careful preparation for the hunt required knowledge of the prey. It had to anticipate the animal's well known habits. The primitive form of reasoning applied here to one specific individual was based on the hunter's general knowledge of the species. If the animal's behavior was characteristic, the hunters were reassured of the approach to take and moved in for the kill. This mental process would later be called deductive reasoning.

Coupled with these mental practices of prehistoric man, another factor may have contributed to his mental development. It is believed that early man's erect posture nurtured development of his brain. Possibly erect posture, walking and running created a greater demand for blood to be supplied to the heart and brain. At the same time, locomotion on land also called for greater vigilance than did the relative safety of arboreal life. Perhaps on the ground he felt greater stress, requiring continual attention to the surroundings, more sustained alertness. Walking exposed him to dangers as voracious animals, poisonous insects and snakes, toxic plants and other primates.

Human consciousness must have undergone a quantum leap due to the investigation and exploration of the immediate landscape so as to anticipate, avoid or ward off any predator or enemy. In this way, the environment offered new, challenging and threatening experiences. It would seem that the move from tree to ground may have conditioned men's skills of perception and focused corresponding mental faculties.

Under primeval conditions, to learn was to survive. In terms of evolution, the processes inherent in the cellular, somatic, and organic makeup of *Homo erectus* were probably activated to such a degree that they actualized the further growth of the brain. More importantly, this stimulation probably led to the further consolidation and integration of the activities of the cerebrum, cerebellum, medulla, and spine. The brain's convolutions most probably indicate that its morphogenetic/symbiotic capacity was increased exponentially. That meant it burgeoned in complexity and intricacy; correspondingly, it compounded skills in perception and comprehension.

Let us now put these speculations in the context of our theory. It is believed that two universal processes govern and evolve all animate nature and the emergence of human intelligence. In the context of our earlier description of morphogenesis, as a process inherent in nature and humankind, man's rudimentary inductive exploration of the environment and the creative ingenuity making tools, weapons, clothing, shelter, and using fire seem to manifest the use of his morphogenetic intelligence.

On the other hand, whenever early man drew conclusions based on elementary induction, he was actually exercising his symbiotic intelligence. When the mind interconnected information in order to integrate the solution to a problem or to utilize an opportunity, there we witness symbiotic purpose. Hence when an integrated plan of action is executed, there we see symbiotic intelligence at work.

Survival Intelligence in Animal and Man

Survival intelligence is manifest in nature, animal and man. The first form of survival intelligence is readily identified. Metamorphosis among animals is nature's way of ensuring a greater probability of survival from the early stages of development to full maturity so that creatures become able to feed themselves and reproduce their own kind.

In the human being, with physical change comes transformation of character and understanding. As the human child grows from weakness to strength in adulthood, from helplessness to self-reliance and responsibility, human consciousness acquires survival intelligence.

As the other animals in nature, we too learned the skill of camouflage detection. It was the ability to discover what was hidden in the surrounding. Gradually, it became focused as in-depth perception. Over time, as humankind learned to practice introspection, the skill enabled us to identify our own self-doubts and confusions as if they were concealed adversaries. This self-exploration led not only to a consciousness of one's inner being, but also to the development of moral conscience.

Morphogenetic/Symbiotic Intelligence

The second form of intelligence prominent in nature and man is the morphogenetic/symbiotic. For purposes of clarity of definition, we may discuss individually these corollary stages of a single, integral intelligence.

Morphogenesis creates and develops cells, bodily tissues and organs. In the human mind, morphogenesis accounts for our gift of imagination. During periods when our skills in survival were tested, we underwent an apprenticeship in ingenuity and resourcefulness, which eventually evolved into conscious creativity. As if in compensation for physical weakness as an animal, mental morphogenesis, coupled with the determination to survive, motivated us to multiply and transform our ways of thinking well beyond the survival intelligence of animal kin. In sum, morphogenesis in the human being became enduring natural curiosity about our surroundings, the habit of exploration and experimentation, and our creative thinking.

On the other hand, in nature symbiosis is defined as the living together in intimate association or tolerant mutualism of dissimilar organisms or creatures. It is a fact of nature that symbiosis between species makes possible their mutual survival. Among human races, it enabled families, tribes, and societies to live side by side in relative harmony. Peace and mutual tolerance most assuredly contributed to their mutual survival.

Correspondingly, within the individual life form, its tissues, muscles, organs, metabolism, and life support systems function together with the mutual purpose of sustaining life. It is the symbiosis intrinsic to the human body itself that has been largely responsible for the integration of human intelligence over thousands of millennia.

Moreover, symbiosis accounted for humanity's intelligent relationship to Mother Earth. Early we learned to cooperate with, obey and even worship her benign abundance. Since earliest man, we must have considered ourselves her children, and every plant and creature had its place and purpose. Through the intuition of Mother Nature's benevolence, our symbiotic intelligence grew. Gradually, we came to sense that nature's symbiotic oneness had a mysterious purpose of its own.

However, the history of medicine has taught us that our tissues, organs, and nerves form an intricate nexus, which not only functions perfectly as a symbiotic system but also probably accounts for the integrity of our body-mind. What once we thought was our vital spirit or soul, we now understand as a life process that integrates us into a being. It gives us the sense that we have some manifest purpose as individuals and perhaps as a species.

So it appears that symbiotic intelligence has been in existence ever since the beginning of distinct forms of life. Over millions of years, cells became "chains of being," which shaped themselves in turn into mutually acting organisms that finally evolved into the vital bioprocesses that sustain our life.

Otherwise, symbiotic intelligence is the essential quality of the human mind. Humankind's intrinsic symbiosis is the source of our perception, conception, and successful cerebration.

Furthermore, the purpose of symbiotic intelligence is to coalesce personal experiences into meaning, knowledge, and wisdom. Indeed, our inborn symbiosis is the reason we define what we learn and know. Symbiotic purpose is what would give any epistemology, based on nature, any hope of knowing why we exist as intelligent beings. Along with morphogenetic creativity, symbiotic intelligence is at the heart of civilization and culture.

Obviously symbiosis alone does not account for how human intelligence came to be. Morphogenesis was always the necessary, complementary life process. In as much as we change physically and mentally throughout a lifetime, morphogenesis and symbiosis act together to effect our physical development and our mental integration. It would appear self-evident that, in animating generation upon generation, these interacting processes are largely responsible for the evolution of humankind.

As more and more past civilizations and cultures are explored, human history provides evidence that over the millennia man has been extending and evolving the dimensions of human intelligence. As we ourselves know, study for its own sake and the use of imagination for the pleasure it gives bring deep gratification. Through such activity, we discover delight in ingenuity and the consolation that comes from integrating life's experiences into homespun wisdom. However modest be the result, morphogenetic creativity and symbiotic consolidation of experience show human intelligence at its best.

Morphogenetic and Symbiotic Modes of Reasoning

Is it possible that forms of reasoning used over hundreds or thousands of years show morphogenetic and symbiotic characteristics? For instance, is it possible that our inductive reasoning is predominantly morphogenetic whereas deductive reasoning is principally symbiotic?

Our inductive reasoning is probably a result of the morphogenetic exploration of the environment for camouflaged forms and movements. Based on past experiences, our symbiotic instinct for the pattern of creature activities enabled us to identify different species. Gradually, this instinct and mode of morphogenetic detection further educated our symbiotic perception to discover universal design, correspondences, and meaning.

Put another way, deductive reasoning is primarily based on past experience. When humankind encountered new creatures or met the unexpected, memory of past events enabled us to apply that experience. We used it as a premise to test the never-before identified or understood. If it corresponded to that experience, the new verified our knowledge. Such confirmation was basically symbiotic.

On the other hand, continued exploration of the environment for further evidence is morphogenetic. This search had the purpose of verifying the value of any symbiotic premise previously formulated. The reciprocal nature of the interaction between morphogenetic exploration and symbiotic confirmation and consolidation of knowledge is based not only on the polarity inherent in nature but in human nature as well. Indeed, this polar reciprocity characterizes the modern psyche/mind itself. Our conscience continually reexamines our earlier assumptions to verify their validty.

History, religion, culture and civilization provide evidence of this fundamental trait in human character. Often literature presents reversals of self-understanding, of world views and men's fortunes. These reversals frequently lead to fresh insights into human nature and humanity. In other words, our morphogenetic creativity and symbiotic conscience continuously test and verify experience. Where survival demands it, these natural endowments can overcome noxious attitudes and transcend pernicious convictions. Hence humankind educates itself by its creative and moral insights into life's experiences.

In biogenetic terms, through evolution and its trial-and-error "method" of dealing with life-threatening conditions or confrontations, the human race has learned creatively to reverse its limited viewpoints or surpass its innocence and ignorance.

In the context of my interpretation of evolution, when tentative or final conclusions are reached, that is the moment when symbiosis makes its presence evident. At that point, the ongoing dialectic of morphogenesis and symbiosis returns to a kind of homeostatic pause, condition, or state.

More specifically, if morphogenesis presents a thesis, symbiosis may offer an antithesis by way of testing or objecting to the soundness of the thesis. By contrast, when symbiosis cautions or contradicts a morphogenetic thesis, there can ensue a synthesis between them, eliciting the best or most viable resolution to the opposed arguments. This synthesis will be a higher or more complete symbiosis and may be conclusive at the time.

In short, the interaction of inductive and deductive reasoning to test and verify the validity of its opposite form of reasoning is likely due to the evolution of the human mind via the morphogerietic/symbiotic dialectic over time.

Morphogenetic Experience, Symbiotic Verification

As a life-form, we have been created by the process of cellular morphogenesis and designed by somatic symbiosis. Thus the function of each bioprocess is to check and balance the other. Our symbiotic intelligence has the responsibility of confirming the truth of what morphogenesis has discovered. This is obvious in our rational need to verify whether or not an experience is true in itself. On the other hand, symbiosis checks if that experience corresponds to memory, i.e., to experience consolidated into significance.

Moreover, when a fact is true to itself, the reason is that our world is an integrated symbiosis whether analyzed by physics, chemistry, or biology. In addition, we ourselves are an integral part of the physical, chemical, and biological world. So truth per se must be true to all spheres of our experience. All true knowledge and wisdom must be consistent with the universal Symbiosis that designed and perfected our evolved intelligence.

We are microorganisms living in the macroorganism called earth. The Renaissance considered man as a microcosm in the macrocosm of the universe. Either way, our inborn creativity and our natural conscience must finally check and balance each other, for seeking universal truth is our main hope of survival for ourselves and our planet.

Nature as Source of Human Imagination

It may be that human imagination is primarily an outgrowth of nature's protean potentials. We have seen how morphogenesis has the power to create., mutate and transform. Yet its counterpart, symbiosis, has the power to coordinate, synchronize, and coalesce nature's most elemental energies into life forms that survive. In view of such creations, it should be evident that mutual interaction represents not only a form of natural intelligence but also a mode of supraconscious imagination.

Once the primeval powers were activated, there emerged plants and creatures never before seen on the earth. In order to survive, these life forms evinced resourcefulness and ways of dealing with life-and-death problems that exercised both their inherent intelligence and imagination. As potentials became processes, there emerged a purpose in nature to design individuals and species able to survive and to recreate their own kind. Replication gave way to innovation, mutation to variation, and finally to evolution. From the beginning of life on earth, intelligence and imagination in some form were present in all that lived.

This contemporary knowledge of nature starkly alters the nineteenth century interpretation of evolution. For the Darwinians, the history of biology was the record of the evolution of species that left behind failures and capitalized on successes. Yet, in fact, evolution was not solely a ruthless struggle, a history of mutual extinction. Today evolution can be explained as the outcome of nature's creative experimentation, tested and perfected over time by nature's symbiotic conscience. How otherwise to explain the countless, ingenious, beautiful, delicate and majestic forms of life on earth today?

In the context of human history, success has meant the manifestation of imaginative intelligence mentored by an ethical, cultural conscience. Thus came about the creative and moral evolution of the human mind.

It would seem that the creation of all living forms (e.g., fungi, plants, invertebrates, chordate, prevertebrates, vertebrates, mammals and Homo sapiens) must be due to the interaction of morphogenesis and symbiosis not only in a given biospace but also across vast eons of time. This morphogenetic/symbiotic interactualization may be interpreted as a dialectic inherent in evolution.

Ergo, in Homo sapiens the dialectical process of intelligence and imagination is manifest in our extensive knowledge and inborn wisdom as to the sense and meaning of our life on earth. Human intelligence and imagination are largely due to the evolution of our specific morphogenetic/symbiotic nature.

However, it is one thing to call mankind Homo sapiens; it is quite another to explain how he knows or how he thinks. What is the biological source of his knowledge and modes of reasoning? Does he have an innate epistemology due to his genes and senses; his evolutionary and ontogenetic heritage; or his morphogenetic/symbiotic experience of existence? How does his imaginative creativity complement and transcend what he thought he knew?

In part, knowing implies being able to fit together myriad sense data synesthetically as well as to integrate existential experiences into a sound, common sense coherence. On the other hand, drawn from widely separate areas of life, imagination fits together multiple data, real and imaginary. Sometimes these become for us universal "laws" of nature and sometimes such archetypal patterns of experience become cultural wisdom.

The Evolutionary Increase in Human Intelligence

Since in higher animals cells specialize, it would seem that any integrated organism is guided by some kind of "foreknowledge" as to the outcome of the morphogenetic/symbiotic process animating its life.

Put another way, morphogenetic specialization and differentiation of cells is counterbalanced by the symbiotic coordination and homeostatic consolidation of cells. Moreover, at times this interactive process continues in multiorganisms till they become superorganisms. In the human body itself, cells join into tissues, cartilage, bone, muscle, blood, metabolism, and life support systems. Thus, our own body as an entity became capable of remarkable somatic-cerebral skills and integration. This by reason of our morphogenetic/symbiotic composition.

In the evolutionary context, the continued interaction of these two processes would seem to promise greater mental capacities for Homo sapiens in the future. This speculation becomes more feasible in the light of the surprising growth of the human skull in a few million years. Possibly there has been a corresponding increase in the brain's neurological growth and connections through the stimulation of human intelligence over the past 100,000 generations of humankind.

The realization that individual intelligence can increase over a single lifetime due to exposure to a wider range and variety of experiences justifies such speculation. Hence an older person should be "smarter" by reason of accumulated knowledge and wisdom as to the sense of life's experiences.

In conclusion, the further evolution of our body-mind could actually be accelerated if we consciously nurtured our morphogenetic and symbiotic endowments. At all levels of education, we would need to teach generations to come that these natural processes need to be matured in privacy and promoted in our systems of formal education.

Simplification and Complexification in Nature

In nature, the processes of simplification and complexification seem intrinsic to all phases and stages of natural growth. The single cell may multiply and grow in complexity to the point when such complexification must be restrained and controlled, or through excessive proliferation (as in the case of tumors), it will do great harm to the host organism.

Indeed, restraint and control seem the primary function of symbiosis in the early stages, which regulates and organizes the activity of cells so that they become aptly suited to performing their own, particular, life-sustaining function or are better prepared to achieve a higher order of growth.

Otherwise, unrestrained morphogenesis would continue its pursuit of complexification if its excesses did not prompt a necessary reaction from symbiosis. In response, symbiosis modifies or curbs morphogenetic activity by a process of simplification which serves the purpose of establishing greater efficiency in the functioning of cellular or organic growth.

Once simplification has tempered growth, renewed morphogenetic creativity will occur, somehow now aware it has a more intricate, more effective goal to achieve. To be sure, that growth will again involve complexification seeking ways and means to attain greater viability and survivability. Again at a check point, symbiosis will be actuated to verify, test and evaluate the accomplishment thus far. And so it will go from tissue to organ to life support system.

At each stage of interaction between morphogenetic experimentation and symbiotic concentration, a higher and more intricate embodiment will take place. With every further simplification via their coordination, cooperation, and integration there will evolve a plant, animal, or human being better able to survive the contingencies and unexpected challenges to staying alive.

Our Mental Processes

As with the processes of simplification and complexification in nature, our mental processes may well follow a parallel pattern. For instance, the activity of linguistic analysis is an attempt to simplify the complexity of language so that we can understand the basic elements of sound, sign, and significance of the single utterance. However, if left at that stage of analysis, language study could disintegrate speech and writing to the point it might become an apparent chaos, that is, the total disintegration of any linguistic meaning at all. In sum, left at that stage of analysis, language would devolve to the grunts, groans, whines and whimpers of animals or our prehistoric ancestors.

Thus one must ask how far can descriptive linguistics or deconstructionism go without a sensible reconstruction?

Philosophically speaking, how long can we continue to emphasize the riddles, paradoxes, antinomies, the unanswerables in modern man's world view? That road leads to despair, cynicism, or fatalism. It mocks man's mind and his capacity to understand what surpasses our present day knowledge. Pure analysis pursued single-mindedly to the point of distintegration represents a real hazard to cultural sanity. It is comparable to the psychic disintegration of individuals studied in psychoanalysis. On the other hand, some scientists in search of a theory of nature practice reductionism or oversimplification. In oversimplifying the complexity of nature, they risk hastening to pseudo-integrations, pseudo-world systems, or even to proposing pseudo-sciences. This becomes evident when an explanation of an array of phenomena proves to be skewed or inadequate because of serious omissions. Subtle but significant facts may have escaped notice because of the limited knowledge of the age. Or one's hypothesis is overvalued. Or a failure of imagination may miss the ultimate significance of the scientific facts. In many otherwise methodically trained scientists, the latter may be the most common defect.

As an antidote to such shortcomings, scientists might undertake to establish a kind of scientific calculus, which is, generally speaking, a system or arrangement of intricate and interrelated parts. Such a calculus would be kin to integral calculus, aiming at the integration of symbiotic parts, functions, and interactions among segments of the truth. Such consolidation would exclude nothing essential to our understanding of the complexity of the subject under study, and yet would include everything relevant so that complete knowledge could be attained by holistic research.

Thus similar to the psychic integration of the sound mind, the rational processes should combine, coalesce, and consolidate all pertinent information so as to make possible a greater symbiosis of sense and signification, of the empirical and noetic, and of the phenomenological and ontological. Such integration would emulate the ancient Greek motto "a sound mind in a sound body." Only here it would imply "a sound theory in a sound body of knowledge."

However, such intact interpretation can come about only where there is authentic interpretation. While the intuitive leap or profound insight is to be cherished in its own right, all too often intuition may remain only that. The reason is that any man's ignorance is far vaster than his knowledge, and any man's knowledge is limited to his particular civilization to educate that intelligence. If the educational system is woefully fragmented, as in the contemporary world, or where it is sadly distorted by reason of religious fanaticism or philosophical inflexibility, then interpretation must fail in so far as the cultural vision is myopic or astygmatic.

To overcome such fragmentation of vision, the sciences, humanities and human sciences must eventually organize worldwide to eliminate effete methods of study and to publish openly all breakthroughs achieved around the world so that mankind's total knowledge may be pooled. Such united purpose would enable the human race to attain a measure of universal knowledge and a more acute understanding of humanity's capacity to master its destiny through mutual wisdom.

6

FORMS OF CULTURAL, RATIONAL AND NATURAL INTELLIGENCE

Since the ancient Greeks, the best minds of Europe and America have occupied themselves with defining human intelligence in weighty tomes. So to devote a single chapter to the topic is to reduce a complex and encyclopedic subject to the brevity of a tentative sketch. That is the intent of this chapter.

The following conspectus will briefly provide a general definition of intelligence followed by descriptions of cultural rational, biological and philosophical intelligence.

Generally, a dictionary would define intelligence as either the ability to learn and understand the new or the capacity to deal with unexpected, vexing and difficult situations. On the other hand, to learn is to discover the facts and truth of a situation so as to determine the connections between the facts and their meaning. However, in order to define the meaning of a situation, we need to discover the reasons it seems so difficult, vexing or inextricable. Hence, as in life, we need to learn how to learn and to understand how our natural intelligence leads us to decide what meaning our life is to have. In essence, that maxim expresses the point of this chapter and this book.

Cultural Intelligence

The earliest evidence of human intelligence, known to the layman, is expressed in Scripture and religious literature. For the West the Holy Bible is the source of Judaism and Christianity. It is said to be based on divine revelation centered on God's justice, love, and the promise of salvation.

The Old Testament contains 39 books and the New Testament 27. The Old Testament includes psalms, proverbs, narratives, and histories whereas the New Testament is made up of four versions of the New Testament written by Matthew, Mark, Luke, and John along with Paul's epistles to the Corinthians. Perhaps the ancient Hebrews depicted in the Old Testament may be characterized as imbued with a strictness of conscience. By contrast, the New Testament was inspired by a more charitable, creative conscience in its promise of a life after the sorrow and suffering of this world. It also obliged the faithful to offer aid and compassion for all humanity.

Over the centuries, the Bible has undergone many translations from Hebrew to Aramic, to Greek, Latin (the Vulgate) English, German, French, Italian, Spanish and 1400 other languages. Obviously every translation involved a dimension of literal truth but also underwent, inadvertently or intentionally, many distinct interpretations from one language to another.

Similarly, the Holy Qu'ran (Koran) is regarded by Muslims as Allah's (God's) actual words revealed to the prophet Muhammad in the seventh century A.D. Its poetic and exalted language embraces laws, moral precepts, and narratives in 14 sutras or chapters. It stresses conviction there is only one God, and that is Allah. Moreover, Muhammad is God's messenger. The faithful must pray to Allah five times a day, practice charity, and lead morally pure lives. Self-discipline and surrender to the will of Allah are considered ultimate virtues.

However, the religious form of intelligence is to be distinguished from an even earlier form, that of the mythological. Myths distinguish themselves in two ways. Traditionally, they were stories about gods and heroes who affirmed a credo based on a heroic view of life. Usually, myths took place in some timeless past. Some myths attempted to explain imaginatively the mystery of existence, the creation of the world and the gods, the reason for life and death.

In later centuries, folklore bore witness to the mores and traditions of medieval society, often believing in the power of the magical and miraculous. Sagas retold the amazing exploits of culture heroes whereas legends often illustrated saintly lives that gave life a moral meaning. The striking resemblance of their motifs and themes show humanity's fascination with such stories over a wide range of cultures. As such, they are thought to express creative imagination through universal archetypes or demonstrate reciprocal, cultural exchanges throughout the world.

Nevertheless, religious intelligence is more seriously concerned with defining moral thought and behavior. It is intent on obeying the will of the Almighty. Life is to be led according to some higher devotion or purpose so that our souls might be acceptable to God in the after life.

At the same time, it should be noted that exalted language is often used to illuminate scriptural passages with an aura of supernatural inspiration. Hence the religious sense of wonder at the ineffable mystery of life is frequently expressed by the analogical or metaphorical mind. This mystical outlook on existence shows humankind endowed with the capacity to experience the ecstatic and to consolidate epiphanies into spiritually meaningful insights.

In antiquity, in contrast to the mystery cults of the Middle East, the Greeks expressed a quite opposite form of intelligence in their use of irony. The usual use of irony is to make a statement the opposite to its literal meaning. As used in ancient Greek tragedy, there emerged a tragic incongruity between the actions undertaken by a hero and the fearful consequences of that undertaking. Thus the result of a sequence of actions was the reverse of what the protagonist strove to achieve.

Hence tragic irony occurs when blindness to the effect of one's own actions (*hamartia*) or pride (*hybris*) leads an individual to make rash decisions or blindly pursue plans that result in his own death. In the fifth and fourth centuries B.C.E., the tragedies of Aeschylus, Sophocles, and Euripides especially come to mind.

On the other hand, one can practice ironic intelligence by pretending to be ignorant and by assuming an attitude of willingness to learn from another. However, the humble posture is used in order to elicit answers from the other which make clear the other's inadequate knowledge or outright ignorance. The ancient Greek philosopher Socrates used this form of irony to question politely the know-it-alls of his time.

Over time, sophisticated individuals came to use irony to test someone's conscience by using a tone of voice to alert the listener to reflect on intolerant attitudes or to revise thoughtless opinions. Today an ironic situation occurs when the self-assured aspire to a position for which they are quite unfit or seek a success they are quite unprepared to achieve.

In another way, irony distinguishes between superficial and in-depth knowledge. In naturalistic terms, irony is basically camouflage detection, distinguishing between appearance and reality, existence and essence, pretense and actuality.

Moreover, over time, humankind has discovered through social interaction with the self-centered, selfish, proud, vain, opinionated, and self-satisfied that nothing educates such individuals more effectively than the use of appropriate irony. One sharp jab of irony usually explodes the ballooned ego.

Otherwise, ironic intelligence acts as a counterweight to any excesses of metaphoric intelligence. The honest, well-balanced mind not only avoids unconscionable irony against the innocent and defenseless. It also restrains any extravagant use of analogy or metaphor, as practiced by demagogues, to inflame prejudices and passions against law-abiding people of other beliefs, races, or religions. The sound-minded individual seeks to harmonize these two traditional attitudes towards life and humanity.

Otherwise, metaphoric intelligence serves a serious, higher purpose in seeking to connect in meaningful ways the facts of the visible world and the invisible truths within us. Moreover, metaphoric intelligence developed into a mode of deciphering truths not apparent at the literal level of meaning. In other words, beneath the exoteric significance of religious and secular literature, there could be an esoteric message.

The polar characteristic of human understanding leads us to consider the role interpretative intelligence has played. We know from human history that the major religions as Judaism, Christianity and Islam have experienced schisms in their temples, churches and mosques. These schisms were the result largely of divisive arguments among members about the significance of holy scripture; at other times, interpretations were "heretical"; and when skepticism set in during the nineteenth and twentieth centuries, there came outright rejection of holy writ as witnessed in agnosticism and atheism.

As a consequence, this conflict of views led interpretive intelligence eventually to some unexpectedly fine insights into "revealed truth" of Scripture. Over history, Biblical hermeneutics developed more carefully worked out methods of investigation and developed more thoroughly substantiated interpretations. Actually these intuitive differences and more sophisticated scholarly studies led to more universal truths about the world's major religions.

Even in the scientific age, these heuristic skills in interpretation helped us evolve more thorough and exacting theoretical explanations of the physical world and the universe. Yet like the antinomies which have cursed and blessed the history of philosophy, conflicting interpretations served in the long run to hone our interpretive intelligence.[1]

Nevertheless, at times we are left to wonder if we actually have knowledge; and when we do, do we have the mental training to interpret that knowledge in such wise as to be assured we have attained a measure of enduring truth?

Let us propose a tentative method by which to ascertain the truth and validity of any interpretation. This basic system suggests how to proceed step by step. First, it would identify a fact as a fact, explain a situation for what it is in actuality, and interpret what it means literally. Second, beyond confirmation of the reality of the situation, interpretive intelligence would use every detail, word and image to make sense of one's own experience of the situation. Third, interpretive intelligence would search beyond and beneath the evident for a larger truth about life. In other words, an authentic interpretation not only would undertake to explain everything clearly and completely. It would also seek out how this experience corresponds to one's own experience of life, whether it confirms or changes one's life philosophy. Ultimately, that is what interpretive intelligence is for.

Rational Intelligence

There are a number of forms of rational intelligence. The most commonly used is comparison and contrast. Comparison matches things, places, persons, actions or ideas that are obviously similar. Contrast distinguishes between them to illustrate or emphasize their differences. Rarely would these rational processes be used in isolation without their complementing the function of the other. Hence when resemblances are noted, so are differences. Similarities are perceived at the same time as dissimilarities.

Together, comparison and contrast function to delineate clearly the relative value, validity, or excellence of two or more things, individuals, activities, or ideas. For example, homes, teachers, sports, or beliefs might be described, portrayed, or set forth with accuracy and in detail.

Comparison and contrast may emphasize the positive and negative, such as indicates the state or condition of a thing. It may be beautiful or ugly, strong or weak, useful or useless, and the like. Yet it is by the interaction of accurate observations and honest evaluations that comparison and contrast provide the basis for rational decisions, judgements, arguments and actions.

Probably humankind acquired early on this habit of mind by learning to detect creatures hidden in the camouflage of grasses and forests. Their dual focus penetrated beneath the surface of things to search the shadows for other forms of life. This bifocal vision in depth likely led them to see through the apparent to the obscured. Of course, when humankind studied the animals in their environment, they soon learned the difference between predator and herbivore and noted the distinguishing marks between related species.

Thus it was the ancient Greek philosopher and zoologist Aristotle who studied specimens of animals to formulate perhaps the earliest definition of definition. Simply stated, a definition is made up of classification + *differentiae*, which, in essence is the patient and careful use of comparison and contrast. In such ways did humankind begin to define their world with greater accuracy and completeness.

Induction and Deduction

Over the centuries, humankind has learned to come to terms with everyday experience in two pervasive ways: inductive and deductive reasoning. Induction infers facts, truths, human ethics, and life experiences. On the other hand, deduction tests and verifies the validity of such inferences by examining new instances or information which either confirms previous conclusions or challenges them by discovering exceptions; or it detects the inadequate logic of the induction.

For instance, it is not an uncommon experience in life to hold a certain conviction as true and worthy of our loyalty only to discover, later, evidence of its shortcomings, harmful implications, impracticality, or actual distortion of the truth. At such times, we reevaluate that "conviction" on the basis of greater knowledge or maturity.

Generally, induction draws its substance and strength from the concrete facts of nature and from the experiences of life, thus leading us to valuable, if tentative conclusions. Conversely, deduction acts to measure the worth of our conclusions so as not to close our minds to the new evidence that nature and life brings us. If induction satisfies our conscience provisionally, deduction intuitively cautions us that we may not yet have all the evidence we need to come to any definitive conclusion.

Note the similarity between classification and induction and between differentiation and deduction.

Problem-Solving Intelligence

Life continuously presents us with problems. Some are truly puzzles or enigmas, but many merely require decisions between alternatives or between mutually exclusive actions. (Obviously, problem-solving not only involves arithmetic, mathematics, and science.) In fact, mundane problems and decisions regarding work, school, family and money are usually solved by the use of common sense and good judgement. In this everyday context, our problem-solving capability is the most universal form of practical intelligence humankind has.

Often the solution to a problem is discovered to be the opposite to the problem itself or the problem turned inside out. Or by going beyond the parameters of the problem, it may be solved in creative new ways. Parameters may simply be unexamined assumptions or old prejudices. They may be due to limited knowledge or the outright ignorance with which we habitually face the unexpected or difficult.

So to solve a problem, the first thing we must learn to do is to examine as completely as possible what the problem is "out there," and secondly, we need to interpret as honestly and dispassionately as possible what the problem may be "in here" in ourselves. After all, the problem may be our attitude, ignorance, arrogance, foolishness, immaturity, vanity or the like.

Conditioned by the circumstances of our social environment, the problem may be due to our blindness to others' needs and feelings. Others may be ignorant about a certain matter, dishonest, willful, or unpredictable. For the most part, if we can resolve inter-human problems by understanding, negotiation, trust, empathy, fairness or old-fashioned generosity, then the practical, hard-headed problems are usually on their way to solution.

In sum, out of such mental habits as: comparison and contrast, classification and *differentiae*, inductive and deductive reasoning, and practical problem-solving, humankind eventually evolved rational methods of knowing ourselves better as well as scientific methods of investigating nature more efficiently. Over the past four centuries, rationality and scientific empiricism gained supremacy as methods perfected in describing and defining knowledge.

Humankind has sought to elaborate comprehensive theories and laws of nature. However, the best minds discovered there could be theories within theories and laws supplanting laws. The absolute understanding of man and the world eluded his scientific mind, despite stunning discoveries of heretofore concealed facts of nature and the development of ever new forms of rationality. Century after century, theories confronted theories, and laws superceded laws.

Forms of Biological Intelligence

Homologous Intelligence

The reader may recall that homology is the study of vertebrate skeletal systems, which reveal the similarity of their bones. These are called homologous structures. They look alike and probably have the same origin in evolution. Such similarities become self-evident when the skeletal structures of different species are compared. For instance, the arm of a man looks very much like the bones of a bat or the wing of a bird, like the flipper of a whale or like the foreleg of a horse.

Perhaps even more striking is the fact that the vertebrate brain, described by cerebrum, cerebellum, medulla, and spinal column, is clearly similar in function between mammal, fish, bird, amphibian, and reptile.

Moreover, microbiologists assure us that homology exists between allelic genes arranged in the same order, and homology also appears in the similarity of amino acid sequences in nucleic acids, peptides, and proteins.

Obviously, such homologous structures and functions are an integral part of evolution and may be said to be an inherent part of humankind's own particular evolution.

When the human mind develops its aptitudes to discover similitudes like patterns of shape and form, phenomena and function, that aptitude could be named the homologous intelligence of humankind. It is this intelligence which enables us not only to detect the homologies among the most diverse species but also to discover the very laws which seem to govern the animate nature we were born into.

A corollary development of mind must have taken place. Pre-historic man learned to use his intelligence to distinguish between shadow and substance as a precautionary strategy to detect the dangerous from the harmless or the predator from game. This habit of discernment was a consequence of the comparison and contrast needed to define and describe animals and birds, predators and prey according to some form of reasonable identification. Thus over time the skill in homological definition arose by man's classifying and differentiating creatures.

Thus, preoccupied in camouflage detection with survival and the education of our perceptions, human intelligence evolved a homological method of identifying life-forms: plants, insects, reptiles, amphibians and the like.

Holistic Intelligence

Moreover, this habit of homological classification had to be based on two interacting mental processes: analysis, which divided a whole into its parts, and correspondingly, a form of synthesis, which reintegrated parts back into wholes. Obviously this aptitude came from early man's skill in cutting up animals to cook and eat them. This "hands-on" activity gave him, as it were, lessons in animal anatomy. Eventually, this awareness of the interconnectedness of a carcass probably led to further skills in developing handicrafts, building shelters, making tools and weapons, each serving a specific use. Over generations, he became aware of other physical connections as between cause and effect and between forces in nature and their consequences.

Thus humanity's holistic intelligence evolved by eventually coming to understand that animate nature was made up of vast, invisible forces that interacted with their own purpose. All nature seemed an immense living organism. Hence in contrast to his earlier fascination with parts and details, there emerged a larger understanding of nature's wholeness. This reversal of vision encouraged humans to systematize their knowledge of the natural world.

Hence was born the insight that existence might have a holistic explanation. Somehow, the world was more than the sum of its parts.

Homeostatic Intelligence

As the reader probably remembers, homeostasis describes a cell's or an organism's stable internal environment. It is a condition necessary to maintain life. The steady state is, nevertheless, capable of producing and using energy. Most important is the fact that it carries on its life functions in an integrated manner, which, in turn, maintains a balanced, internal environment.

Now the fact is that every living creature survives by homeostasis, including ourselves. As the equilibrium that maintains the distinct but interdependent elements in a single cell or in an entire organism, the human mind itself seeks such homeostasis in the balance, harmony, and integration of all its needs, desires, aptitudes, and experiences, those induced by our senses and those generated by our innermost being.

That means the mind needs to heed its own checks and balances so as to maintain optimum mental health. The mind needs as much knowledge of the body as of the outer world through which it moves.

Body and mind need to respect each other by coming to terms with their combined homeostatic nature. By so doing, body and mind together represent our homeostatic intelligence.

Morphogenetic/Symbiotic Intelligence

Homeostasis reflects nature's most pervasive process, symbiosis. As the reader well knows, symbiosis generally means the intimate living together of two dissimilar organisms in a mutually beneficial relationship. Moreover, this active, interactualizing relationship characterizes the essential nature of every living body.

Nature provides us with an abundance of evidence that symbiosis is active at all levels of energy exchange between individual life-forms, and it accounts as well for the mutualism between species. Likewise, each and every creature, from the simplest to the most complex, survives by reason of the symbiosis intrinsic to its own physical nature.

It should be self-evident that all life-forms must use some form or degree of symbiotic intelligence. However, such intelligence is displayed not simply through the mutual interaction of life support systems within any given life-form. In more evolved species, symbiotic intelligence is also incarnate in their neural and cerebral consciousness of what it needs to survive.

At however elementary a level of consciousness, all life forms must take decisive actions or perish because of their inability to do so. In nature, symbiosis means knowledge of the prospects of life and death in the environment. Thus one reason for species extinction may be the failure of species to develop sufficient symbiotic intelligence to survive.

Just as symbiosis is ubiquitous in every life-form so is morphogenesis universal. Even the simplest cell is able to initiate changes in form from within. Self-transformations serve the cell's single-minded purpose of survival. On the other hand, in multicellular organisms, the purpose is not only to stay alive but also to increase its prospects of survival by better organization and more effective integration.

This process may be due to trial and error experimentation with its own nature to discover whatever strengthens its chances of survival. To be sure, the process would also involve more obvious experiments in adaptation to various environments.

In more evolved organisms, biologists have found that distinct species in different natural media share homologous structures. Hence creatures of land, sea and air adapted their body shape and body structure to the medium they live in. Such transformation must primarily have been initiated by morphogenesis from within.

Biologists have argued that evolution may proceed by slow variations or by sudden saltations. In other words, there is evidence that morphogenetic intelligence usually goes on at a steady pace. Otherwise, given favorable conditions or faced by challenges threatening its survival, morphogenetic intelligence may actually accelerate its own evolution.

Now in the case of human evolution, nature's trial-and-error experimentation appears to have been primarily mental. Early on in our history, imaginative individuals may have already practiced a form of introspection so as to discover who they were and what they wanted out of life. Since there were periods when humankind was less preoccupied with survival, those were times free to pursue our natural curiosity, to innovate, to invent, to create. It is probable that humanity's morphogenetic intelligence underwent a marked acceleration by reason of the fact that, through human culture and civilization, humankind initiated a coevolution to that of nature.

Mankind's ingenuity and resourcefulness mark us as a species apart. Even our mental meanderings may be more than idle daydreams. As imagined experiences they may be mental experiments. They may explore memories, intuitions, and conscience to discover ways to mutate into a still higher form of intelligence. Or maybe our cerebral curiosity is seeking to grasp the meaning of our lives.

Ontogenetic Intelligence

The term ontogenesis refers to the development of an individual organism as manifest by visible morphological transformations. As human beings pass from birth through childhood through adolescence into adulthood, middle and old age, ontogenetic change is represented.

Our personal experience of life teaches us to view our lives with a greater understanding of the meaning of time. One of the invaluable consequences of growing older is that we mature our ontogenetic intelligence, which maturation comes only with time. Whatever wisdom humanity has accumulated over the millennia is due principally to this particular manifestation of intelligence.

Time puts things into perspective. Time helps us evaluate the worthwhile from the less valuable. Time helps us understand our failures and successes, our defeats and victories. It registers the days of happiness and those of tragedy, the hours of fulfillment and those of loss. Time matures our ontogenetic intelligence and makes our conscience more just, sophisticated, and compassionate. It better grasps the meaning and purpose of life.

Evolutionary Intelligence

Solid evidence of mankind's evolutionary intelligence is harder to come by. If we focus on the paradoxes, antinomies, and conflicts Western civilization has been heir to, one might be left with a feeling of intense skepticism and cynicism as to any evolution of human intelligence.

Indeed, there is much evidence to gainsay any optimism. Violence in the streets and the home, strife between races and religions, industries poisoning the good earth all are hardly cause for assurance of the future. Epidemics arising from human filth and ignorance, and mayham, murder, and war in every part of the globe all question whether humankind manifests even the intelligence of the so-called "dumb" animals.

Philosophical Intelligence

When we consider the rise and fall of civilizations, the flowering and withering of cultures , we sense there have been, over time, manifestations of evolved intelligence. We remember the great minds of the ancient past: Moses, Christ, Siddhartha, Gautama Buddha, Socrates, Plato, Aristotle, Pythagoras, and Plotinus.

We remember the names from the Middle Ages—St. Augustine, St. Thomas Aquinas, Mohammed, Dante. We recollect the greats of the Renaissance as A. Dürer, Leonardo da Vinci, Michelangelo. We recall the composers, artists, scientists and philosophers of the sixteenth through nineteenth century, such as J.S. Bach, W.A. Mozart, Voltaire, Rembrandt van Rijn, Galileo, Copernicus, Sir Isaac Newton, L. Beethoven, W. Whitman, L. Pasteur, I. Kant, J.W. Goethe, G.W.F Hegel, Darwin, F.M.Dostoyevsky, L.Tolstoy; et al.

Our collective memory holds these geniuses and other unforgettable ancestors in justified veneration. They illustrate in how many ways human intelligence can be developed. They teach us how far future mankind and womankind can evolve with intelligence worthy of our species.

Consider the history of ideas over the past three millennia. Key ideas have influenced the minds of countless generations. Consider such concepts as: chance versus fate; free will versus predestination; the question of God's goodness and omnipotence in the face of evil and human suffering; sin and salvation; vice and virtue; wickedness and wisdom. In different times and cultures, such extremes tempted, afflicted and tested the "souls" of men and women, and by so doing developed what civilizations called moral conscience.

Over the ages, we have witnessed how human intelligence varies and transforms, not unlike the evolution of life-forms in nature. Thus, in general, it would seem that ideas evolve and survive if they are intelligent or go extinct if they are not.

Indeed, life itself clearly manifests virtually endless emanations of form, design, pattern, color, texture, function, survivability, and perfection. Ever since the first bacteria and cells appeared on earth, the billion year heritage of living things has been with us and acts through us as a peculiar species capable of thinking with an evolved intelligence.

A Tentative Summary

The modes of understanding bequeathed us by nature are the true source of humankind's natural conscience. Obviously this view of human nature transcends the orthodox, censorious conscience that some practitioners hold up as the moral model for all humanity.

Our evolved intelligence recognizes how the archetypial metaphor "the great chain of being" retraces the evolution of human thought. That symbiotic interpretation of existence inspired some 500 generations of western men and women to seek some greater meaning to their lives. It inspired the hope that humanity and the universe had a common bond. If true, it would be possible that human life, indeed, had some eternal significance.

Moreover, this germinal inspiration guided the philosophical education of western civilization. During twenty-five centuries, the great minds sought to express the faith that existence is, in fact, a chain of infinite being and meaning.

Thus, each great mind, remembered and studied, provides us with solid evidence of the evolution of human intelligence. For as surely as we study: (1) the record of geology, (2) fossils in paleontology, (3) comparative anatomy, (4) parallels among species in comparative embryology, and (5) the development of the simian/human cranium over time, we find in the products of humanity's greatest minds encyclopedic evidence that the human species has been undergoing some sort of great mental evolution over the ages.

Hence in so far as we devote time to studying the artistic, religious, literary, philosophical, scientific and technical contributions of the past and present, the teacher and student of this and future centuries emulate the cultural evolution of the human race.

Today we still can take refuge in the measure of hope that knowledge of the past brings us. For that knowledge is the most convincing evidence that human intelligence has actually increased over the million years before and the millennia since the beginning of recorded history.

The mental development of our ancestors, forefathers, and ourselves is undeniable testimony how far human nature has transformed morphogenetically and symbiotically. Modern man has an evolved intelligence. Nature urges us to use that inherent creativity and conscience so as to perfect them in the millennia to come.

7

HOW NATURE'S PROCESSES CREATED HUMAN INTELLIGENCE

Introduction

In the centuries immediately preceding the Third Millennium, the sciences studied nature "objectively" as the source of universal truths and laws. The primary method of studying the external world was empiricism, the use of our physical senses to observe, test and verify the verities of this world. This earthly orientation had its own rationale. It turned away from "the revealed truth" of religion because it assumed the actual world contained the ultimate truths of existence.

In this context, Darwin's theory of evolution showed deference to the concepts of "causality" and "mechanisms."

Apparently he felt obliged to imply such means of effecting change in the evolution of species. This terminology tarnished somewhat his brilliant and comprehensive presentation. Otherwise, his painstaking research and thoughtful conclusions significantly extended our vision of life on earth.

The interpretation of existence as a "struggle for survival," where "the fittest" pass their evolved strengths to their descendants, greatly enhanced our understanding of the evolutionary meaning of animate nature. Unfortunately, in another respect, his interpretation led to a life philosophy akin to the Old Testament in so far as his implicit biological determinism evoked a sense of despair that we have very little control over our lives. The individual seems a hapless victim in the face of the inexorable forces in nature.

Consequently, Darwin's *Origin* left his generation and those that immediately followed with a sense of fatalism. The law of predation and the life-and-death struggle seemed to justify the law of the strong and ruthless. The frail note of hope that all violent victories in nature were sometimes necessary for the sake of biological progression made a mockery of traditional morality. The social consequence seemed that the individual could be sacrificed in the name of "progress" for human civilization.

Of course, such a philosophy of evolution would justify wars and man's inhumanity to man because events confirmed the fundamental laws of nature. After all, superior tribes, nations, races, and species simply were obeying their "manifest destiny."

In the last half of the nineteenth century and throughout the twentieth, biology directed its study inward to the microbiological and to the processes intrinsic to all life forms. Internal processes were recognized as mutually functioning together. In addition, microbiologists uncovered the intricate essence of the natural world. This innermost reality existed within every life-form on earth, including humankind.

Slowly, but surely, biologists and philosophers perceived that this internal world gave us an understanding of evolutionary processes distinct from Darwin's interpretation. It was time to look within for answers.

As we became aware of the inner environment that all living creatures carry within them, we began to realize that our inner world required microspection as well as introspection. We began to study humankind as biological beings with millions of years of evolutionary experience in our cells, genes, life support systems, and brains. All these had been directing our responses to life and educating our intelligence as a species.

Today, biologists are beginning to formulate a new understanding of life. Existence is not only what we see with our eyes—the plants, trees, forests, deserts, prairies, mountains and all the creatures of land, sea and air. Rather, nature is in all living things. It is their inner world. As regards humanity, that hidden reality makes us who we are and what we may become. Nature is in every part of our body. The processes which actuate all forms of earthly life course through our own body and mind.

An early philosopher of evolution, John Henry Green (1791–1863), professor of surgery, believed that nature was "…striving toward greater individuality and consciousness."[1] (Richards, 78-79)

Green's statement prompts us to ask how and when consciousness might have arisen in cellular or organic life. Upon reflection, it would appear self-evident that a life form becomes conscious when it "realizes" it is a being, an entity, an identity. Eventually over time a being must also become aware of its own becoming. So consciousness may be said to occur the moment a cell, organism or life form senses it is alive. In evolved mammals that "moment of truth" or consciousness probably occurs at the moment of birth.

This preliminary speculation calls for a further query. Is it possible that subsequent life forms—cellular, genetic, organic, somatic—might each represent stages in the evolution of self-consciousness? Such self-consciousness would become more than an awareness of being alive in a given environment.

Darwin supposed that each of us, during embryogenesis, comes to "pass through the evolutionary history of our species." (Richards, 97) If so, it may be possible that our body remembers its bio-history. Our genes may act as our evolutionary database to process our new experiences. But exactly what are instincts, intuitions, insights, and consciousness?

It would appear that our bioheritage brings with it a sense of the purposiveness inherent in all animate life-forms. We know now that evolved forms and designs of nature each serve a purpose of its own. It should be obvious that human nature also incarnates such purpose. At our evolved level, would it not be to attain true conscience by realizing the essential purpose of our life.

Let it be made immediately clear that conscience extends beyond any civilized notion of sin or guilt. While those notions most assuredly have their moral worth, nature is not moral in the sense of religious ethics. If anything, nature's morality goes beyond manmade commandments or righteous dogmata, however worthy they may be or however valuable in fostering civilization.

In the context of biological nature, conscience does evince one salient feature kin to our usual ideas of morality. Human conscience implies mutualism, which means mutual aid among mankind, that is, the readiness to help others survive.

Conscience is manifest in some form throughout humanity. In human beings it originated and developed in a special way through apperception guiding perception. In other words, old experience evaluates the new. Obviously, over thousands of generations, this process went through stages of maturation. Gradually mature understanding developed consciousness into conscience. Of course, maturity and conscience came about by the individual coming to terms with the demands of the environment and with the increasing complexities of life. If the individual survives into adulthood, conscience will provide worthwhile insights and understanding, and these principles will guide conscience far into the future.

Now genuine understanding comes about when one learns to make a generalization about particular instances in life. Put another way, when the conscience learns to simplify the complex into a simpler, clearer comprehension, it has learned to replicate a process that occurs all through evolution. In order for nature's designs and forms to function efficiently, it must simplify the complex so as to enable the life-form to survive. In terms of human nature, conscience guides the individual's life purpose.

Along the way in living life, the individual pursues a variety of experiences. This pursuit duplicates the natural process called *divergence* which in humankind urges us to identify our true needs and extend our interests so as to enhance our prospects for survival. On the other hand, the individual also learns to draw on the corollary natural process of *convergence*, which guides us to concentrate our life energies so as to pursue worthwhile goals.

Such concentration teaches conscience the benefit of self-control, temperance, and self-discipline. Concentration and conscience undertake to harmonize one's intelligence, talents, and needs so as to survive and to achieve life-fulfilling objectives.

Moreover, the human mind has developed more than consciousness and conscience. Life's variety and challenges have had their benefit. They have taught us the need for resourcefulness and creativity not only to survive but also to create new possibilities in life well beyond basic survival. Eventually, such resourcefulness leads to all the inventions man has ever invented, leads to all the practical methods of agriculture and technology, our advances in science and medicine.

In combination with consciousness ever alert for the significance of experiences, creativity has created new knowledge. In combination with conscience, creativity has created the laws, institutions, forms of government, order, morality, and ultimately all the actions and gestures of man's humanity to man.

Our discussion thus far has given us four insights. First, although the scientific study of "visible" nature has taught us much about its phenomena, the microbiological study of "invisible" nature has found the universal processes that animate all of nature. Second, the new science of the microscopically small seems to have discovered a degree of consciousness in all forms of life: the cell, the gene, tissues, organs, life support systems—all performing their functions perfectly as it were consciously. Third, in the evolution of vertebrata, if we consider the maternal and paternal instinct of mammals and hominidae, there appears evidence of the growth of natural conscience. Fourth, beyond consciousness and conscience, the human mind has developed a remarkable capacity for creativity.

These introductory remarks now lead us to consider, in some breadth and depth, how universal processes in nature have conditioned the evolution of the human mind. It is nature's polarities which are responsible for the actualization of human nature. This fact will be demonstrated by showing how divergence and convergence, complexity and simplification, morphogenesis and symbiosis interacted to evolve our body and mind. Thus due to the dialectic of these polar processes, intelligence emerged in nature, and the human mind became our biogenetic destiny.

Universal Processes in Nature

Embryonic Development
and Lifetime Transformation

A human embryo starts life from a single cell, itself a symbiosis of sperm and ovum. This cell proliferates through stages: growth into multicellular organizations as well as differentiation into specific tissues, organs, and life support systems. These developments serve an immediate, mutual purpose, the survival of the life form.

Birth, childhood, adolescence, adulthood, middle age, old age all manifest stages of maturation and, finally, physical and mental decline.

Corollary to the immediate and ultimate mission of the embryo, at every stage the baby and child discover that it has specific needs. On the other hand, it usually takes decades for the adolescent and adult to think of questioning the reason for life. Ignorance of our biological heritage accounts for this lack of interest.

Nevertheless, the development of the human embryo is a clue to the past and a foreshadowing of the future. The embryo must somehow sense it will go through successive stages of life. Each of these stages serve a purpose, which requires its consummation before beginning the next stage.

The embryo offers us a lesson about life. As each stage of embryonic development requires us to develop specific tissues, organs and somatic systems, so too in a lifetime we need to develop specific skills in order to survive successfully. Later stages of life do not require us to develop new organs; instead we need to acquire new forms of knowledge. These will enable us to function with more understanding. In all its efficient and effective forms, knowledge makes us stronger and more able to perceive how life-enhancing activities evolve intelligence.

The human embryo offers another lesson in life. If all its life support systems exist to execute particular functions, the completed embryo has one compelling purpose—to be born. In the perspective of a lifetime, that necessity would seem to indicate your own life must have a destiny, but not simply to survive as an animal to gratify its biological needs. Humankind are given life to find their truest identity, to learn the purpose and meaning of our one opportunity to live.

At natural stages of life, we should take into account the facts, conditions and experiences of our existence: (1) Where are we in terms of self-discipline and accomplishment? (2) What worthwhile ideals do we live by? (3) What honorable objectives are we seriously pursuing? (4) What does life mean to us at this time? (5) At the end, what will we have done with that one life?

Maturation means coming to terms with our need for emotional fulfillment. It also attests to compassionate concern for others. Maturity signifies we have completed stages of learning about life's significance, for we ourselves have given life its meaning.

Such an accounting should help us realize certain truths. We ourselves are responsible for initiating inner changes in attitude and understanding. This initiation is a process of self-education. Changes are required, if we are to live in this ever changing world with intelligence, sensitivity, and humanity. Our self-transforming civilization requires the individual to learn to live peacefully with billions of other humankind who share the earth with us. Once we acknowledge this fundamental necessity, we are ready for the next stage of human maturity.

Human Evolution

If there is any truth to the theory that the human embryo recapitulates the evolution of our ancestors' adult forms, then perhaps our genes, instincts, and psyches have a faint memory of their experiences. Although this suggestion seems improbable, we do have knowledge of our ancestors in other ways. At least, most families have a rudimentary recognition of names in a family tree, even though the names may mean nothing to those alive. Yet the name of each family member represents a life lived and a human destiny.

Beyond our own lineage, we all have a world full of noteworthy and memorable ancestors from every region of the earth, who have reached forward to us in time—their descendants. Our true ancestors are all those who have left behind records of their lives divulging their thoughts, feelings, hopes, despairs, their accomplishments and defeats, their life plans. Our true ancestors are all the architects, engineers, scientists, inventors, philosophers, doctors, artists, writers, and composers who contributed to our understanding that humanity is composed of individuals with intelligence, conscience and compassion, people who lived their lives to share their strengths, knowledge, sympathies, and wisdom with each other.

Put another way, our ancestors speak to us every time we open a book or read a page from the past. They speak to us: (1) when we look at a fine building or a magnificent monument; (2) when we learn a scientific theory or marvel at an invention; (3) when we appreciate a work of art, read a literary masterpiece, or listen to a musical composition. Our ancestor's voices are all around us every time we visit a library, a museum, a university. And those voices whisper to us across the ages from every culture and civilization that humanity has given birth to.

Yet, as individuals, we can show respect for our flesh-and blood ancestors. How? By remembering that because of them we are here, alive. If we are not to prove ourselves ungenerous and ungrateful, we need to take time to think of who they were and what they did for us.

If we are honest, we will ask ourselves some hard questions. Was all their hard work and self-sacrifice wasted? Was their undeniable love for us, their hope for our future wasted? Did their devotion to their children and their tender love for their children's children simply come to mean nothing?

Let us put it another way. How are we living our own lives? Do we have the right to throw our lives away in senseless activities and trivial occupations? Do we spend our days in meanness of spirit and our nights in loveless sexuality? Do we show disrespect for others? Do we denigrate what our forefathers held in honor and respect?

Such questions should prompt us to think of the ultimate meaning of human evolution.

When we learn that as a species we have evolved through stages Cro-Magnon, Neanderthal, Homo erectus, Homo sapiens, we may ponder anew the meaning of human evolution. It may flatter the vain to imagine we are intelligent, clever Homo sapiens at the apparent acme of the evolutionary curve, but what about the next stage of evolution for humankind, if there ever is to be one?

"Idle conjecture" someone might say. Perhaps. Perhaps not. If we look to the natural processes by which humankind have evolved, we may obtain clues or hints as to the direction and goal of our future evolution. At the very least, both embryonic development and lifetime self-transformation show us that humankind can evolve as individuals and as a species mainly by pursuing a lifetime commitment. But which natural processes can serve to orient us in that direction?

There are six processes in nature which have enabled us to evolve. These six intersecting lines of force help generate and integrate the body/mind of the human being. In turn, these entelechies have served to evolve the human species through superior stages of life and intelligence. These six archetypal processes are: divergence and convergence; complexity and simplicity; morphogenesis and symbiosis.

Polarities Actualize the Evolution of the Body-Mind

Divergence

The most common understanding of the term *divergence* is "the evolutionary tendency or process by which animals or plants that are descended from a common ancestor evolve into different forms when living under different conditions."[2] (Barnhart, 171) Furthermore, Julian Huxley states "Through the process of divergence each species exploits the resources of the environment more effectively, so that the large scale result of divergence in the inhabitants of a region is comparable to the physiological division of labor in an individual body." (Barnhart, 171) These definitions provide us with a viewpoint largely governed by the scientific study of outer reality. Yet Huxley's analogy offers a clue as to the hidden process of divergence governed by the inner reality of a life-form.

Any divergence must be initiated from within. Divergence must be the process which generates the multiplicity of new multicellular tissues and organs. This enables the organism to create the distinct properties needed to meet various living conditions. Indeed, divergence is one conspicuous way that an organism can adapt to distinct and new environments. The organism does so by changing the environment within to accommodate itself symbiotically to the demands of the exterior.

Obviously if nature provides the biological means by which to initiate somatic divergence, then a similar capacity should characterize the mind. The process of mental divergence is seen in the way the human mind has learned to explore and create many different methods of perception, apperception and reasoning. (To be sure, human history has shown how humankind have already evolved multiple modes of reasoning, imagining, intuiting, and introspection.) So it seems this instinctive capacity of mind arises from the very nature of our cells, genes, organs and soma.

It is necessary to make clear how these evolved faculties and skills are a consequence of nature's universal processes, which account for our species' evolution. Furthermore, we need to examine how the polarities of evolution (divergence and convergence, complexity and simplicity, morphogenesis and symbiosis) provide insight into the body-mind's inherent functions. If this can be ascertained, then perhaps it may become possible to schematize these processes into a comprehensive system of perception and apperception. Such a system would be of use to the natural and human sciences, to all kinds of creative thinking, to the humanities, and to more efficient and effective education.

Thus divergence is evidently the process by which we extend, broaden, and universalize our discovery of new knowledge, thereby increasing our potential for further mental achievement. For the more we know, the better we can make connections between our experiences, the better we can draw sound conclusions as to what we have learned. So by acquiring insight into diverse fields of knowledge, we should ideally and actually enhance our comprehension, intelligence, knowhow, and foresight because we have grown aware of a vaster spectrum of opportunities for the future.

Convergence

The term *convergence* is generally understood as "...the act of converging and especially moving toward union, or uniformity; especially, coordinated movement of the two eyes so that the image of a single point is formed on corresponding retinal areas." (Tenth, 253) The coordination of eye movements is a clue to the coordination of the entire body. A glance at human anatomy and its obvious symmetries provides evidence of the evolution of somatic convergence in virtually all known life-forms.

This initial description is confirmed by the scientific definition of the verb *converge*. In biology, it is "the tendency in organisms to develop similar characteristics when living under the same conditions." (Barnhart, 128) Furthermore, convergent evolution "is the appearance of similar characteristics in organisms not closely related to one another." (Barnhart, 128–129)

These definitions point to a process underlying the evolution of species. (Wainwright in *Axis and Circumference* offers a similar, convincing argument that the cylindrical shape is ubiquitous throughout nature.) Such uniformity of form, such similarities of shape (whether homologous or analogous) clearly provide evidence of the omnipresent process of convergence actualizing the designs of animate life. Wherever there is symmetric physiology, convergence has helped realize it. Even where there are asymmetrical cells, tissues and organs within, since they make up networks of life support systems, such intricate asymmetries have united to sustain the life of the life-form. As such, they too are the consequence of convergence.

Just as convergence integrates cells into multicellular organizations and eventually into complete organisms, a similar process must take place in the brain, psyche, and mind.

In the case of the human brain, it should be self-evident that it is the result of the integration of all cells, organs, senses, and systems of the human body. To be sure, integration is part of the task of somatic convergence.

In the case of psyche, it should be similarly self-evident that it is the outgrowth and ingrowth of all the somatic, sensory, sentient and noetic experiences of the life form we are.

The term "mind" is rejected by a number of scientists and philosophers. It is a complex concept which attempts to account for the mental and emotional characteristics possessed by human beings. Among these traits are disposition, desire, memory, will, intellectual ability and conscious mental events. To describe how these diverse properties are coordinated and integrated, thinkers use the abstract term *mind*. It is a philosophical generalization about a whole range of mental phenomena. Hence the concept itself is the result of a conceptual convergence.

The process becomes clearer when we consider how human beings learn. Through formal and informal education, through apprenticeship, through life's experiences, we practice convergence of our impressions, ideas and thoughts. Convergence takes place when we concentrate all our energies or mental faculties to solve a problem, to complete a task, to achieve a goal. When we learn to fully concentrate the mind on a single task, we are on the way to self-mastery. Through steady practice we acquire patient self-discipline to dedicate hours, days, weeks, months, even years to a single problem, goal or pursuit.

Obviously men of the greatest achievements have developed this capacity and ability to concentrate, and by so doing, they replicate both the patience of nature itself and the convergence of intelligence that humankind have inherited from our evolutionary past.

As explained below, divergence is the result of universal morphogenesis; convergence is the result of universal symbiosis.

Furthermore, it is important to note that the capacity of divergence and convergence is shared by all humanity, for these processes largely account for human history.

Growth of Complexity Versus Simplification

Toward Complexity

Another universal process in nature is the growth of complexity. In the development of a human embryo, for instance, the growth of cells, tissue, and organs results in a great intricacy of somatic and neural organization. This growth in complexity is accompanied by an extension of its capacities and functions. Ultimately, the growth results in complete life support systems and a perfectly formed baby.

To be sure, during an individual's lifetime this growth of complexity continues, but more concentrated in the development of the human mind. Advanced studies in medicine tell us that the brain is the center of the body's network of senses and life-sustaining systems. In fact, the brain orchestrates and directs all sensory responses, neurological reactions, and physical actions.

Moreover, the brain is the organ of neural coordination, capable of interpreting and correlating stimuli, capable of thought. As the center of higher consciousness, it is the nerve center of human intelligence.

As the nexus of the body-mind, how does it manifest its own complexity and its capacity for greater growth of complexity? Obviously, by the mind's ability to know and to create, it reveals itself. It is able to invent (e.g. Edison's electric light bulb), to work out systems of thinking (e.g., mathematics), to describe laws of nature (e.g., Newton's *Principia*), and to conceive the nature of the universe (e.g., Einstein's theory of relativity). Furthermore, it is able to create great works of music and philosophy.

In all these accomplishments, we witness not only their own intrinsic complexity. We come to realize that they have enhanced the growth of human knowledge and nurtured the mind's own growth by extending our comprehension of the intricacy of existence itself. Finally, complex concepts tend to incorporate knowledge into spheres of universal significance.

At the very least, intricate ideas enrich our estimation of the powers of the human mind itself and proclaim the value of all mental and creative work.

Granted that complex ideas may be of little interest to the man or woman weighed down by the cares, concerns, and worries of everyday survival, yet these great, complex ideas tend to prove humankind have a purpose in the universe. We will know it once we have learned the meaning of human destiny.

Simplification

The process of simplification is also manifest in biological nature. The reason is that limitless proliferation of complexity could suffocate a life form. Just as the individual plant can be strangled by the luxuriant encroachment of its competitors, so too can the single form strangle on its own, undisciplined inner growth. Thus nature initiates a limit to proliferation either by exuding toxins to ward off aggressors, or it simplifies the inner complexity intrinsic to the individual. Hence to check, control, and delimit complexity, there appears a point at which the growth toward complexity abates and a counter phase is introduced. Structures, processes and functions are simplified for the purpose of economizing the flow of energy and to ensure that everything works with greater efficiency.

Simplification also serves a purpose in our mental processes. Ideas, imagery, symbols and patterns of reasoning increase the effective power and definition of our perceptions, thoughts, and argument.

Moreover, simplification concentrates the meaning of impressions, data and phenomena. Simplification focuses greater efficiency by perceiving the essence of a problem or enigma. (e.g. Oedipus solving the riddle of the sphinx).[3] Furthermore, simplification keeps the mind from being entrapped in a labyrinth of deadend passageways. Rather, simplification traces out mentally the most direct path to escape from confusion and entrapment. In addition, it is able to collate and condense widely separated fields of knowledge and insights so that a complete, holistic explanation can be extricated.

As we learn from Freud's psychoanalysis, Jung's analytical psychology, and the contemplation of world literature, the study of imagery, metaphor and symbol can help us see through the complexity of human suffering to draw valuable life lessons from such study. Simplification enables us to understand the pathological patterns of the possessed psyche and the tragic consequences in the lives of the rash and reckless, the unwary and unwise. In other words, the intelligent use of simplification allows meaning to emerge. It enables sensitive, intelligent human beings to come to terms with their lives.

In expository writing, simplification establishes clarity, order, and unity to our insights and intuitions. It defines our knowledge and proves our argument. Through simplification, we gain greater mastery over our thought processes, we plan our activities more efficiently and communicate our ideas more effectively.

Morphogenesis

As our earlier discussion made clear, morphogenesis performs two major functions. (1) By activating cellular activity, which initiates the generation of multicellular organisms, it fosters growth within life-forms. (2) By exploring the environment in which it finds itself, morphogenesis pursues a correlative purpose.

Thus morphogenesis is active in two complementary phases. In the first, it explores the scope and limits of the external environment because the habitat provides food and opportunities for survival and reproduction. In the second, as a consequence of experience gained from exploration of the outer world, the creature would at some time need to explore its own inner world. Obviously, success would reinforce the survival strategies it used. Failures would require the life form to find out how and why it failed.

To the thoughtful reader, these two phases sound familiar. The process of exploration we might find to be a kind of inductive reasoning. On the other hand, the need to test the effectiveness of one's plans, thoughts and actions we might identify as a kind of deductive reasoning. Hence we seem to discern familiar mental processes in these two phases of morphogenesis.

Experimentation might be another function of morphogenesis. As practiced in chemistry, the unknown may be tested by two procedures. As is done in smelting metals from ore, one procedure may be to extract an essence from coarse matter. Another procedure is to distill the impure so as to leave a pure and useful residue.

Nature itself provides us with parallel examples. The roots of a plant extract nourishment from the soil, avoiding the toxic and ignoring the inert. In the case of an animal, its digestive system acts to neutralize the harmful, to remove the indigestible, to assimilate the nourishing, and to evacuate the rest. In their own way, these natural activities in plant and animal seem the consequence of nature's morphogenetic experimentation.

Moreover, morphogenesis also functions to transform natural elements into food. For instance, the main task of plants is to transform rainwater, soil minerals, and sunlight into food for itself. On the other hand, in animals, mastication of coarse edibles and through their digestion by gastric juices converts all to useful nutrition. Furthermore, the digestive system enables the body to change these raw materials into nutrients for flesh, bone, organ and brain. Nature's morphogenetic experimentation acts to transform whatever in the environment can be of use to the survival of the life form.

Experimentation also characterizes one way the mind functions. Part of the brain's inductive accumulation of experience and its deductive conclusions may be said to be experiments in gathering knowledge. Such reasoning is not only the domain of scientific thinking or research. In fact, all human experience in contact with the environment is empirical—the survival service of our senses.

Moreover, every time we ordinary humans test our sensory experience by applying it to a new situation, condition or event, we clearly are using a form of deductive reasoning. Together, induction and deduction constitute, in essence, experimentation with life's experiences. Put another way, our perception becomes apperception by transforming our observation into thought and understanding.

Discovery is another characteristic of morphogenesis, for it searches the environment to ascertain what is real, fact and true. This action is applied not only to the outer world but also to the inner world of the life form itself. Indeed, one might say that all life is a process of discovery. If the habit first develops to anticipate immediate survival situations, eventually it becomes a lifetime purpose to understand the meaning of one's discoveries. In this respect, no life form is fated. Rather, pursuing a lifetime of discovery implies the individual is seeking to find out what life means and what one needs to survive. Over time, this habit, leading to viable decisions, would evolve into determining one's own destiny.

Selectivity is another function of morphogenesis. Darwin's emphasis on natural selection is not to be forgotten. Implicit in the term *selection* is the conviction that there is adequate, innate intelligence in life forms to be capable of choosing between mutually exclusive alternatives, of avoiding circumstances that threaten their lives and adopting strategies that ensure survival. In other words, animals are not fated in their biological destinies. To be sure, all must die, but the species live on, sometimes over millions of years. Over eons of time, successful species seem to have made enough right decisions or have developed enough intelligence to overcome whatever threats, injuries, or evils fate may have inflicted upon them.

Creativity is a final characteristic of morphogenesis by its capacity to create new possibilities, new alternatives, and new prospects in one's life. From the smallest cell in our body to the most intricate network of living matter, morphogenetic creativity is deeply imbedded in our intrinsic nature.

It is highly probable that morphogenesis provides the vigor and vitality of embryonic development. Indeed, it is the mysterious power made evident in birth. Yet beyond bringing the newborn into the world, it continues to energize the growth and maturation of the human being through an entire lifetime. Beyond the single life, morphogenesis animates the self-transcending experimentation in all evolved beings. In other words, morphogenetic creativity fosters the further evolution of its most successful experiments so that they endure as species over the ages.

In sum, if morphogenesis displays traits as exploration, selectivity, experimentation, discovery, and creativity, such functions are evidence that morphogenesis is a manifestation of intelligence in nature. It is a fact that the earth offers abundant proof that a superior intelligence permeates all of animate nature.

As to human intelligence, ours is only one form that nature evolved. In the light of the evidence presented thus far, it should be obvious that the time has come for human kind to consciously promote morphogenesis in our private activities as well as in formal education. We need to educate the young and ourselves to emulate the abilities nature has evolved over the eons. If we truly are the children of mother earth, it is time we appreciate her ageless skills and wisdom.

Symbiosis

Morphogenesis does not act alone. As seen in the natural processes already discussed (divergence and convergence, complexation and simplification), nature acts by checks and balances. In the proliferation of life forms, we saw how divergence was controlled and guided by convergence. Whereas divergence generates a multiplicity of cells, tissues, or organs, convergence integrates those cells, tissues and organs into multicellular organisms and into life-support systems. If divergence brings forth life through various stages, convergence organizes, concentrates, and integrates it into wholes, networks, and plans.

Similarly, whereas cells and tissues grow in complexity to develop ever more intricate, "intelligent" organisms, such growth is countered periodically by the process of simplification, which checks, controls and delimits complexity so as to bring about greater efficiency of biological functions and greater economy of energy. By this interactive process, nature evolved organisms with more effective means of auto-control (homeostasis). In addition, this dialectical process established the body's life-support systems networks of intra-communication to deal with survival situations.

In like manner, we find that morphogenesis is monitored and managed by symbiosis, its polar partner. As previously noted, when morphogenesis grows quiescent, symbiosis takes over. Thus the purpose of symbiosis seems to be threefold: to coordinate, to consolidate, and to integrate cells, tissues, organs and systems. It should be obvious that divergence or diversification of a life form has to be delimited just as the growth in complexity has to be simplified so the life form can better survive. Symbiosis, is the powerhouse behind convergence, which moderates limitless differentiation and is behind simplification, which restructures and "streamlines" complexity. Thus symbiosis periodically supervises morphogenetic processes so as to assure their viability and durable value.

If this is so, what does symbiosis teach us about the brain's inherent properties and about the way the mind works? Symbiosis acts at various levels in the human being. First, through our senses, symbiosis unites our perceptions and impressions synesthetically so we are able to synthesize our experiences into more mature understandings. Second, at given moments or stages of life, symbiosis requires us to come to terms with our intuitions, intimate emotions, and practical knowledge of life's requirements. This integrative need helps to educate us to survive in intelligent, sophisticated ways. Moreover, symbiosis helps us, as symbionts, to define our essential identity among humanity. For that identification will allow us to establish valuable, reciprocal relations with others, and, in addition, urge us to achieve a worthwhile destiny of our own.

In general, how can we awaken people to have faith that life has a purpose? How can we inspire the individual to pursue a destiny worthy of all the sacrifice, love, hard work, and intelligence of our forefathers? How can we ourselves serve humanity?

There are certain actions we can take. First, we need to recognize the deeper meanings to Darwinian evolution as amended by contemporary biology, ecology, and microbiology. Second, we need to learn the importance of time in our private lives and as the sum and substance of our civilization and culture. In-depth knowledge of humanity would require us to study, with respect and open-mindedness, the religious, philosophical, intellectual, scientific, and humanistic contributions of our ancestors, who bequeathed to us their understandings of the significance of human destiny. Nor should we neglect our world myths and folklore or our heretics, skeptics, and mystics. Third, we need to educate our descendants to appreciate the actual values of our own age, not only the discoveries of sciences and the inventions of technology but also the creations of the humanities.

Moreover, in as much as the evolution of the species has provided the biological basis for the evolution of the human intelligence, we need to understand how our mental processes replicate that natural evolution. If the above description of divergence and convergence, growth of complexity and simplification, morphogenesis and symbiosis is a fair account of the innermost processes powering the evolution of the species, then we may undertake to educate the modern mind accordingly. By teaching the conscious and creative use of these inborn processes, we prepare future thinkers to evolve an authentic epistemology in accord with our ageless biological heritage.

Tentative Consclusions

Evolution

1. The human mind is the issue or progeny of biological evolution.

2. At birth, our embryo becomes a being. Over a lifetime, the being evolves a thinking mind with a human conscience.

3. Human adaptation means we must learn to adapt ourselves to outer nature and to our inner being of thought and feeling.

4. Survival requires we respect our bodies and minds.

5. If we wish to study the stages of actual human evolution, we need to grasp the significance of our ancestors' achievements and accomplishments.

6. The record is there for all to read: their mores and customs; their religions and ethics; their architecture, craftsmanship and engineering; their sciences and technology; and their music, art, literature and philosophy. All these document the creativity and conscience that evolved humankind.

Mutualism

1. In nature, mutualism between distinct forms of life means they aid in one another's survival whether intentionally or inadvertently.

2. Mutualism also means that members of the same species have been known to defend, support and comfort those needing it.

3. In humankind, mutualism means uniting in a common purpose to meet the needs of human desperation, suffering and anguish from whatever cause. Mutualism means combating poverty, starvation, and disease. It means providing relief and succor wherever war and natural disaster have turned a human being into a victim. It means befriending all humanity, regardless of nationality, race or religion.

Divergence

1. In nature, divergence and diversification lead a species to establish a special econiche so as to better assure its survival.

2. To the human mind, divergence strongly suggests that we discover our special identity, talent and capacity for creating a life.

3. Divergence also suggests that we develop a lifetime specialization, which makes secure our own survival and protects those in need of us.

Convergence

1. In nature, convergence means the unification of our cellular, organic, somatic, cerebral being into a viable life form.

2. In the human mind, convergence strongly suggests the need to develop mental concentration and to integrate our creativity, knowledge, experience, and intelligence.

3. Convergence also tells us our lifetime purpose is to unite all that we are into a destiny worthy of our aptitudes, talents, character and comprehension.

4. Finally, convergence suggests the establishment of relationships between those who need each other. Such union among humankind will ultimately mean we consolidate mind and heart, conscience and compassion.

Morphogenesis

1. Darwin's' theory of evolution argues that nature is a struggle for survival. The "fittest" to survive become those favored by nature to leave descendants. That is what is meant by his expression "natural selection." This ruthless extinction of the "unfit" makes nature appear insensate and indifferent to the fate of her progeny.

2. By contrast, microbiology's discovery of morphogenesis reveals quite a different nature. It permits us to offer a more creative understanding of the intrinsic evolution in all forms of life.

3. The microbiological study of morphogenesis investigates inner nature made up not only of cellular, organic and cerebral processes but of the human mind itself.

4. Our morphogenetic mind makes clear that the innermost urge of our being is to innovate, invent, and transform the imperfect to the perfect.

5. Our morphogenetic intelligence urges us to experiment with life's possibilities and to educate fully all our potentials. Ultimately, morphogenesis provides us with the incentive to create a life identifiably our own.

6. This creative drive may require a lifetime of "soul searching." First and foremost, it may be the search for your innermost psyche or self. It may also be the search for a soulmate. Alone or together, sincerely and earnestly pursued, you will create a meaningful destiny. That is the lesson implicit in morphogenesis. For human beings, that is nature's most fundamental law.

Symbiosis

1. In external nature, symbiosis manifests itself by the intimate association or union of dissimilar organisms or creatures. When relationships are mutually beneficial, they demonstrate the phenomenon of mutualism. Two or more individuals establish a mutual dependance.

2. In microbiological nature, symbiosis is manifest between distinct cells, tissues, organs, and the network of systems that sustain life in whatever form.

3. In the human body, symbiosis merges, coalesces, and consolidates every atom of living matter into its total unity. Skeletal, muscular, nervous, endocrine, respiratory, circulatory, and other life maintaining systems interact and mutually reinforce one another. Thus, body, heart and brain constitute a symbiotic, holistic entity.

4. Obviously both the symbiosis revealed in outer nature and the symbiosis divulged in our inner nature indicate that we, as symbionts, and nature itself are intimately related and mutually dependant. (The most obvious evidence of this is the carbon dioxide-oxygen cycle between animals and plants.)

5. In human terms, the senses orient our instinctive, morphogenetic exploration of the world. The discoveries are intuitively integrated by our body-mind into symbiotic knowledge.

6. Our own symbiosis exhibits the holistic need to coordinate and synchronize the evolutionary polarities of human nature (the propensities to divergence/convergence, complexity/ simplicity, morphogenesis/symbiosis). This integration of opposites clearly is with the purpose of regaining or establishing inner balance and harmony, congruity and oneness.

7. As a consequence, by resolving inner conflicts, the restored accord enables individuals to decide the direction their life should take.

8. At successive stages of life, we formulate tentative symbioses of self understanding which allow us to proceed into the future with matured self-confidence.

9. In the end, pursuing symbiosis lifelong should enable us to integrate our intelligence, knowledge, and comprehension of existence. In this pursuit, we may become masters of our own destinies.

The Significance of Nature's Processes to the Human Mind

1. This chapter has reviewed nature's basic processes, which have conditioned and influenced the evolution of the human body and mind.

2. These natural processes actuate and actualize our being and becoming.

3. Because their mutual purpose is the survival of body and mind, they can be trusted to guide our lifetime decisions and to help us master our own destinies.

4. Educated, disciplined, and perfected, these evolutionary processes can aid the individual to realize the acme of human intelligence.

5. All these processes contribute to our biological cerebral life. In time, they will allow us to grasp the ultimate meaning of a human existence dedicated to the full realization of the intelligence bequeathed us by the womb of life.

PART V

HUMAN EVOLUTION IN CULTURE, CREATIVITY AND CONSCIENCE

8

THE POLARITY OF
HUMAN NATURE IN CULTURE

Introduction

In nature, divergence generally occurs between species, plants, crustaceans, fish, reptiles, birds and animals. Of course, many species reveal homologous structures within (e.g., skeletons), which substantiate Darwin's argument about the origin of species. However, distinct species adapted to different environments often in unique ways. They learned to exist independently and to survive in special eco-niches.

The range of such adaptations not only points to an organic capacity to divergence but also accounts in part for the competition of species for survival. Competition for territory, for food, for mates further accentuated their differences, divisions, and diversification.

On the other hand, convergence is also manifest throughout nature. This term means that creatures living under the same organic or ecological conditions tend to develop similar body shapes and other characteristics of evolution. We have already noted the similarity of body and design between porpoise, shark and fish. There are other striking similarities of other sea creatures as sea stars, sand dollars, sea urchins, and sea cucumbers. The similarity of shape of vertebrate brains may be attributable to a measure of convergence as between mammals, fish, birds, amphibians and reptiles, all of which are distinct species. Finally, the similar environment of the womb most likely accounts for the striking resemblance among vertebrate embryos as that of the chicken, turtle, pig and human being.

This manifestation of divergence and convergence in nature would seem to apply to the history of human nature in culture as well. If we assume that the human brain, psyche and mind embody evolution, then it would appear biological that humankind living in their diverse ecosystems and regions of the world follow their own tendencies to divergence and convergence. We have seen further evidence of the interaction of these processes by the way the mind itself uses them instinctively to refine perception and achieve meaningful conceptions.

Human Divergence and Convergence

Among humankind there early developed instances of divergence as in the case of language. By about 10,000 B.C.E. there probably were several thousand proto languages. These later diversified and divided. A benefit of this linguistic divergence was that a distinct language confirmed communal identity and communication, which enabled families and tribes to live and work with a common purpose. Hence such groups diverged in the direction of ethnic or racial identification. Yet this distinction not only differentiated and distinguished them; to a degree it also separated them from other human groups.

When a proto-language divided into several descendant languages, there arose such distinctness between them that it became difficult, if not impossible, to communicate with people of related languages. The reason was that the mutual etymological roots of their languages were lost in time. Of course, the phrase "Tower of Babylon" is the source of our English word babble to refer to virtually incomprehensible speech.

The phenomenon characterized the break between Japanese and Korean about 2000 B.C.E.[1] This process also occurred during the Roman Empire when the Romance languages diverged from Latin into Italian, French, Spanish, Portuguese, and Romanian.

On the other hand, nothing in life acts solely under the influence of one polarity, for usually one pole is countered or compensated by its opposite polarity. This is illustrated in the acquisition of a distinct language. For instance, English clearly shows that multiple languages have contributed to it. Among them have been Latin, French, German, Spanish, Russian, Arabic and such exotic languages as Japanese, Tagalog, Nahuatl, and Malay. This accumulation of terms from diverse languages means that it continues to accrue knowledge of the world through the enormous variety of indigenous ideas and foreign references. Thus English promotes within itself a great convergence of concepts.

This capacity for convergence of ideas and concepts is especially true today in the major, international languages, which readily assimilate knowledge through the sciences, business, commerce, trade, economics, cognitive systems, methods and key technical, medical and academic advances. In so far as we share such knowledge and know-how, it will undoubtedly help all races, peoples, and religions to better understand and respect each other's intelligence, cultures and civilizations. Evidently this tendency in intellectual and professional language will ultimately influence the convergence of the mind of the human race.

Another example of the interaction between divergence and convergence is the fate of the Jewish people. As prisoners of war in captivity in Babylon (750–612 B.C.E.), the Israelites were deported by the Assyrians to Samaria, where they integrated in a Jewish sect known as Samarians. In Central Asia, they became "the lost tribes of Israel." This diaspora or scattering of the Jewish people is an example of forced divergence and separation yet the Jews in their isolation must have experienced a greater unity of spirit, a convergence of faith and skepticism, of hope and despair. A similar banishment of the Jews from Jerusalem (about 133 A.D.) became another dispersion of the Jews, again an instance of coerced divergence with similar spiritual convergence as a consequence.

Between 656 and 661 A.D., the Muslim religion (Islam) underwent a similar divergence between the orthodox Sunnites and those faithful to Ali as the true Imam/leader of the "sect" called Shiah Muslims or Shiites.

Divergence

In addition, the Inquisition of the Roman Catholic Church sought to be rid of dissenters, disbelievers, and heretics by punishing the slightest deviations in belief with extreme cruelty. To be sure, the outcome resulted in the massive revolt called "the Reformation." Mutual separation and condemnation ensued. Unfortunately, the Protestants also alienated millions by their own form of Inquisition.

A particularly illuminating example was the witchcraft trials in Europe and early America (1560–1600 A.D.) The ostensible purpose of such so-called "trials" was to purify Christians and to ensure complete compliance, convergence and obedience of the "faithful." Of course, a backlash resulted. The excesses of the Inquisition and the witchcraft trials finally came to a crisis of conscience between the blind loyalty of the faithful and their use of common sense and instinct for justice. Finally, this crisis led to wide revulsion and rebellion. Between 1618 and 1648 there raged a Thirty Year War between Protestants and Catholics. The population of parts of Germany was reduced by half due to war and disease.

Divergence between such major factions may have been intended to promote convergence within each group loyal to its own credo, but, in the final analysis, it led to a frightful destruction of life. Human strife to secure ultimate and total agreement failed. It ended in the split between populations who believed in a loving God, Christ. To this day there remains wariness and distrust for the other.

Another form of divergence is also revealing. The astronomer, mathematician and natural philosopher Galileo Galilei (1632–33) was accused and found guilt of heresy by Inquisition judges because he promoted the Copernican view of the solar system. Although he recanted, history leaves us with a lesson from his intellectual humiliation.

In this example we see the suppression of facts and truths that contradict accepted doctrine, dogmata, and acceptable world view. In the guise of infallibility and conformity, and with the purpose of enforcing absolute convergent thinking, "authorities" have historically suffocated differences of opinion and original knowledge. However, ultimately history passes judgement on the "judges."

The whole history of heresy, the punishment and official banishment of dedicated thinkers and fearless leaders led to outrage and clear consequences. Trying to suppress original and intelligent divergence wherever and whenever it arose ultimately failed. The suppression of the innovative investigation of our world or the repression of human talent, intelligence and creativity is bound to fail because the repressors are attempting to annul a law of nature.

Convergence

In contrast to the above examples disclosing the principal effect of divergence on human history, the following cases illustrate the fundamental phenomenon of convergence and its effect on human cultures and civilizations.

One of the earliest examples of convergence was in Mesopotamia about 4,500–3,500 B.C.E. Archeology discovered what appeared a confederation of villages, farmers, traders, and artisans living and working together. Their temple towers, consisting of high, stepped pyramid structures called ziggurats, served religious and secular activities. Temples, government offices, storehouses, and workshops were built around it. (Much as a multicellular organism grows organically and physically, humans seem to construct similarly organized communities to survive.)

In Eurasia, various animals were used to plow the land. In separate places at widely different times, people developed parallel skills, used similar farming and irrigation techniques. Such common practices demonstrate a simple example of convergence.

Throughout history from all over the world, there is abundant evidence that similar skills developed in hunting, fishing, planting crops, and practical problem solving.

Despite the fact that humankind were separated into different races, languages and religions, the fact that diverse peoples achieved similar levels of agricultural, technical and cultural competence is evidence among humanity of mental convergence.

In many places, different human groups were inventing systems of counting. Such cerebral convergence is due to the fact humankind faced similar survival problems. Common ideas, abstractions and the use of numbers began to shape an archetypal mentality.

About 3500–3000 B.C.E., record keeping was begun. The construction of roads, bridges, harbors, flood control projects, canals, and irrigation ditches were the outcome of common efforts. Obviously, such public projects were undertaken to establish (symbiotic) connections, organizations, orderliness, and integration. At the same time, these activities show that the human mind itself was beginning to recognize its own capacity to connect, organize, order, and pursue purposes with practical outcomes. This development appeared in many places around the world in distinct historical periods.

Of special interest is man's invention of alphabets and writing. Around 3400 B.C.E., the Sumerians of Mesopotamia originated their own cuneiform/wedge shaped scripts, believed to be the first writing system devised. This device or alphabet made it possible for words and thoughts to be recorded so as to last over time, in contrast to spoken, evanescent speech. Although speech itself served an immediate, useful purpose, writing eventually served a timeless purpose, that of bringing together individual minds, communities, civilizations, past and present.

The Egyptians developed hieroglyphics ("sacred carvings") as their written language. It combined sounds, signs, and pictures. They also invented the first calendar. Not only are these examples of human ingenuity. Both writing and a calendar served the ambition of preserving knowledge, which in itself is an example of the integration of human intelligence.

In the Indus River Valley of the Himalayas, the Harappan civilization (2600–2000 B.C.E.) developed a written language—part ideographic and part phonetic. In addition to the practical aims of written language, it served the function of enabling a people and a culture to survive oblivion. In brief, a written language could be sent over the earth and across time, such that thousands of years later future readers would know those cultures had once existed.

About 1400 B.C.E., Phoenician scholars established their own famous alphabet. Their superior sound system of 22 signs enabled writers to write down any language. Here again an alphabet encouraged convergence of human thinking and mutual understanding. However, note in this case, the system enabled the Phoenician to translate accurately the sound system of many languages, which became the first step toward universal understanding among humankind.

On the other hand, the Roman alphabet of 23 Latin letters derived from the Etruscan model. The objective in standardizing the spoken language was to be able to communicate written messages accurately. Thus in ancient Rome, they ensured the convergence of command and obedience. Today that ancient written language passes on to us the knowledge and wisdom of their poets, philosophers, writers, and theoreticians, such as Martial, Juvenal, Tacitus, Epictetus, Plutarch, and Lucian.

As further evidence of the ingenuity characteristic of human intelligence in all parts of the world, the classic period of Mayan history and culture (300–900 A.D.) saw the flourishing of mathematics and astronomy, an exceptionally precise calendar, and a writing system based on pictorial and phonetic graphs and signs. It was comparable to those invented in Mesopotamia, Egypt, and China. This is obviously another example of human mental convergence. Where a need became evident, the mind invented a medium to meet that need.

Directly related to the invention of writing systems was the development of translation. For instance, there were the first translations of the Christian bible. Obviously the intent of translation is to convert those who read it. Successful translation leads to the convergence of men's minds and souls to embrace a common faith and to serve a mutual purpose (peace on earth). Through the translation of their sacred works, virtually every major religion has aimed to convert others. Thus this skill also led to a convergence of human understanding.

In another vein, Marsilio Ficino (1433–1499), translated Plato, Plotinus and other ancient writers and philosophers.

Then there was Miles Coverdale (1488–1568) who turned Protestant and published the first translation of the entire Bible into the English language. Here again translation was a means of spiritual convergence.

To be sure, the most memorable work of accurate translation was the work of 51 scholars working seven years to produce the 1610 King James Version of the Bible, which is still the standard version for some churches and denominations.

The importance of translation and scholarship is self-evident. The role of translators and historians seems to be to merge and converge humanity across time. Theirs is an act of faith in human intelligence and the human race. Similarly interpreters of literature, philosophy and the sciences are also in the service of humanity as are all true teachers. The translators, historians and teachers of the mind of humanity act as interpreters of its creativity and moral conscience. These men and women are important to the evolution of human intelligence. They all serve the purpose of guiding humanity to some ultimate, mutual understanding.

A kindred focus of human energy and intelligence has been centers of learning. For instance, in 930 A.D., Cordoba in Spain was the center of Muslim religous, commercial and cultural activities. As a focal point, it integrated the knowledge and faith of the Muslims of the time.

Then again from 1113–1142 the French philosopher and theologian Pierre Abelard helped make Paris a renown center of learning. The importance of learning and teaching is not so much to unite minds to one and only one form of knowledge but to teach mutual respect for one another's learning even in the face of divergent views and the diversity of minds and philosophies. Hence in 1209 Cambridge and Oxford, patterned after the University of Paris, shared faculty, students and curriculum. As a consequence, there developed cultural convergence based on symbiotic, intellectual sympathies.

In 1300–1375 the city state of Florence emerged as a world class cultural center, home of renown European writers as Dante Alighieri (1265–1321), Petrarch (1304–1374), and Boccaccio (1313–1375). Ingenuity and genius were acknowledged, and as most great cultural centers, it converged and focused the best humanistic aspirations of the time.

But we must not forget the Library of Alexandria, famous in antiquity, which had collected much of the knowledge of its time. As prototype for modern institutions with extensive holdings, libraries the world over have done much to help the individual educate himself or herself. Libraries help actualize the education and evolution of humankind. Their collections of books represent the convergence of the best human minds, past and present.

Another example of convergence influencing human evolution is the Egyptian and Mesopotamian standardization of their weights and measures around 3100 B.C.E. to 1100 A.D. Such standardization to prevent the tendency of human nature to cheat others is a move in the right direction. Not only is the intent to adapt to the requirements of trade and commerce but also to establish honesty and fairness in human dealings. Thus it is a move toward ethical convergence.

A similar case is in Harapan sites (800 A.D.) which give evidence of a common system of weights and measures, even arithmetic with decimals. In addition to a central marketplace, there were heated pools and evidence of a covered drainage system.

These archeological facts bear out the thesis that human beings used intelligence in such wise as to promote security, standards, order and a measure of just (economic) relationships. Man's use of signs, symbols and methods to assure exactness and equity could only promote peaceful and honest relationships. Moreover, the use of pools and decent drainage helped ensure a measure of public health. Such archetypal examples of convergence pursuing the common good and well being must have been an important influence in human evolution.

Another source of convergence was architecture in the service of religious ideals and faith. For example, in 691-2 A.D., the oldest surviving mosque in Jerusalem, the Dome of the Rock, was built. Muslims believed Mohammad ascended to paradise there. The third holiest site in the Islamic world, the mosque is a symbol of Islam and Muhammad himself. Since it was the center for the annual pilgrimage (*hadj*), it is a source of spiritual convergence for Muslims. (It is much like the Vatican or Holy City for Catholics).

Going to such holy places renews one's faith and inspires one to pledge anew to one's beliefs. As a source of integration and unity for the believers, their faith helps guide them to trust in Allah/God totally and to pursue a moral destiny.

Christians had a similar devotion. Due to the great period of Gothic architecture in Europe, similar architectural symbols encouraged the convergence of Christians in their daily lives. Prayer, church attendance, annual processions in honor of saints, and periodic pilgrimages gave the faithful a sense of commitment to a higher purpose in life. It was their soul-felt faith that provided them with the dedicated energy, over hundreds of years, to build a number of majestic, imaginative, and awesome cathedrals (e.g., Chartres, Paris, Reims, Amiens). They represent a perfect example of convergence of spirit and technical building skills.

From 1070 to about 1350, a similar fervor and faith inspired the construction of 27 great cathedrals in England.

This and other similar examples point to a significant theme of thought in humanity's cultural evolution. It points to the fervent hope that an ultimate convergence can take place among all humanity—with a common purpose for mankind uniting all individuals, races, and religions into one faith and credo.

An evolutionary influence of a different kind was evinced during the Italian Renaissance (1250–1600 A.D.), which saw a revival of interest in antiquity, its classical art, philosophy, science, its republican spirit and secular humanism. Indeed, the Renaissance oriented humanity to the various capacities of human intelligence through the arts, literature, and philosophy. Initiated by a renewed recognition of human ingenuity and originality, the cultural interests of the age produced a convergence of minds. By the sympathetic study of ancient Greece and Rome, a symbiotic interest in the past was manifest. Hence the spirit and intelligence of antiquity were reborn centuries later. The respect and admiration of the past was another form of spiritual convergence among humankind.

Still another influence on the evolution of the human mind came as a consequence of exploration. For instance, Ferdinand Magellan's circumnavigation of the globe in 1519–1522 had its effect on humanity's understanding of their world. Obviously, such exploration was to encourage an eventual convergence of thinking of the earth as one place, one home for humanity.

Tentative Conclusions

Although history provides many more illustrations of convergence among humankind, it is time to draw some tentative conclusions from these few, representative examples.

1. Even though the earth has multitudes of peoples and religions, our common humanity has the simple, basic, common purpose to survive, to enjoy the world's natural riches, and to live free lives obedient only to a conscience that tells us right from wrong. Humanity has not yet reached such sane and moral convergence, but we may be on our way.

2. Noteworthy in the interaction of races, peoples and religions, divergence often is followed by a more comprehensive convergence as after a war when different philosophies are melded and conquerors and conquered learn from each other. Such melding of factions, peoples, and philosophies are an effective argument against any theory of racial superiority, which is an anthropological myth. On the other hand, based on historical evidence, no sect or faith can lay claim to moral superiority.

    The history of religions provides abundant examples of the consequences of fundamentalism, fanaticism, and self- righteousness. Despite commonly held moral beliefs, their mutual slaughter in religious wars is an indelible fact. Does any side hold the high moral ground when their atrocities against one another speak louder than any mumbled faith in the Creator, conceived as merciful and compassionate? All sides share the guilt and shame of genocide. Hence throughout history, divergence has largely ruled the relationship between the world's religions.

3. We have also seen that education and human evolution are synonymous. Without education in some form, human intelligence can go just so far. During the past seven centuries in Europe, education has meant learning what the best minds have thought, felt and created. This knowledge at least has added to the individual's awareness of the intelligence inherent in humanity. Therefore, the symbiotic aim of education is assimilation and consolidation of humankind's cumulative intelligence, and in so far as possible, to make practical and intelligent use of that knowledge.

   Of course, the morphogenetic side of intelligence requires that we learn the significance of the creativity present in all the great creative works of the past (the humanities). Learning to learn is one thing. Learning to become more intelligent is another. Learning to understand the various capacities of creativity that culture has bequeathed us is still another. Ultimately, to learn what morphogenesis and symbiosis have to teach us about human intelligence is to reach a pinnacle of human evolution we may call wisdom.

   In terms of our biogenetic heritage, wisdom means exercising the Creative Conscience of Nature as it is potential in the mind of man. When the finite individual mind endeavors to converge all its aptitudes, talents and intelligences, it is replicating what has been occurring to humanity across the millennia. When the infinite creative conscience of humankind emulates the immense intelligence of Nature's Creative Conscience, humanity is truly evolving a new creative conscience in the world. By making known what was not known heretofore, by discovering and expressing in our finite terms the infinite intelligence omnipresent in Nature, we begin to sense the presence of a mind in the Universe. How else can we explain the Light and Life we, as a mere species, see and feel within us?

5. Despite diversity in place and time of these various cultures, they provide clear proof that intelligence is a homologous heritage of all "races" and ethnic groups. Expressed through morphogenetic ingenuity and symbiotic consciousness of our mutual mortal destiny, it is such human intelligence that holds the greatest promise of human integration and unity in the Third Millennium. May it become a millennium when all humanity have equal rights, equal education, equal security, equal nourishment, and equal medical aid.

9

THE BEGINNINGS OF HUMAN INGENUITY AND CONSCIENCE

Human Resourcefulness as the Basis of Survival

Approximately 100,000 years ago, man as Homo sapiens hunted and foraged through the Eastern African savanna into the Nile Valley and the Near East.[1] Nature must have been a vast mystery to them. Over and over again it intrigued and lured humankind to explore the earth in every direction. There seemed limitless possibilities full of unpredictable adventures. The longer they survived, the more these experiences seemed to turn into opportunities. By migrating afar and exploring the unknown, they eventually developed resourcefulness and ingenuity. They also learned to expect the unexpected. They learned to live. In time they lived to learn.

In order to survive the unknown, they had to develop skills. They learned to make fire; they made tools; they sewed together crude clothing; they built rustic shelters; they cooked the raw into the edible; they invented language.

They searched the secluded, scanned the remote, scouted the inaccessible. They devised weapons of defense against their predators. They planned hunting strategies to capture game. They learned to plant and harvest crops.

Hence our prehistoric survival required that we use humankind's inborn creativity and conscience. These were gifts bequeathed to us by Nature's own morphogenetic and symbiotic evolution. Instinctively and intuitively, human nature replicated and emulated Nature's ageless powers of creation and integration.

Environmental changes must have further stimulated their intelligence. The seasons and the changing climate with the withdrawal of the Ice Age (50,000 to 15,000 years ago) brought new dangers into their environment (the appearance of new predators) as well as new opportunities (new forms of vegetation and game).

It is believed that during the Old Stone Age (42,000 to 17,000 years ago) humans probably lived in groups in order to forage more efficiently and to more successfully track wild animal herds. Such foraging was a practical means of "living off the land," but it was also the beginning of the "research" of nature not only for the edible but also for the useful and medicinal. Many foragers or searchers meant a wider range of exploration, comparing discoveries, acquiring a more extensive, firsthand knowledge of nature, and utilizing greater knowledge for surer survival. Such activity pursuing a mutual purpose encouraged the development of humankind's symbiotic intelligence. About 32,000 to 14,000 years ago, humans displayed their artistic skills in cave paintings and in the shaping and carving all sorts of utensils and vessels. Indeed, over generations their artisan skills seemed to improve with time. Dexterity in handicrafts showed a natural morphogenetic ingenuity in humankind as well as a practical penchant to use their symbiotic knowledge of nature. The design and durability of many archeological artifacts are evidence of widespread morphogenetic/symbiotic knowhow.

On the other hand, the artists who did clay figures and rock paintings (CroMagnons) displayed both talent and skill in execution. Obviously, the remarkable sketches and paintings of animals, associated with symbolic hunting scenes, demonstrated a keen eye for the real in nature. Yet something else was manifest by their invention of crude symbols representing the unseen like the rocks erected as tombstones or "houses" for the souls of the dead.

Perhaps the natural phenomena of wind and its movement across the landscape made them sense the presence of some invisible power. Or perhaps flowing water seemed invested with a spirit. What mysterious power made a tree or a man grow? The revealed and concealed in the world all around humankind awakened in them a consciousness that there was hidden in him or her a spirit, a soul, or the mysterious thing we call "intelligence."

About 20,000 years ago when the oceans lowered by some 250 ft./130 meters, immense ice sheets covered the Bering Strait 53 mi./85 km.) to form an ice bridge between Asia and the New World. Hence once again environmental change, which opened up new adventures, new horizons, and new opportunities, taught man the advantage of adaptability and exploration. Perhaps the bridge to the other half of the world awakened man's anticipation that life's possibilities were endless.

About 18,000 years ago, large herds of mammoth, bison and reindeer were to be found in open grasslands, woodlands and in river valleys. (King, p. 3)[2] What effect did these animals have on early man? Symbiotic knowledge of their habits and migrations guaranteed him food through harsh times. However, man's ability to carve bones into tools and weapons and hides into clothing and shelter against the severe climate clearly shows his morphogenetic ingenuity.

About 13,000 years ago, the ice began to withdraw. Large mammals started to become extinct. In the far north, new forests appeared and wild grasses covered the highlands. The Sahara was covered by large, shallow lakes and grassland. The oceans and seas rose. (p. 3)

This new growth calls for a question. After centuries submerged and suffocated by continents of glaciers, where did this new life come from? Was it simply carried there as wind or water borne seeds from the warmer climates of the earth? Or were the seed in the earth there since the million year Ice Ages began? Both possibilities should cause us to ponder the astonishing durability and toughness of seeds as well as their capacity to "know" when the right time had come to germinate and where (the right soil nutrients, moisture and duration of sunlight).

Indeed, the simplest seed seems to contain the secret of life and of survival. A simple seed has inborn the capacity of morphogenetic germination and growth as well as the symbiotic capacity to perfect the design of plants and seeds so they could endure prolonged winters. Repeatedly they withstood the inhospitable, harsh climate.

Somewhere from 13,000 to 8,000 years ago, Homo sapiens sapiens were nomads crossing northeastern Asia and slowly settled all parts of the New World, north and south. On the other hand, people also migrated island by island across the Pacific from China, Melanesia, and Polynesia. They left artifacts along the coasts of South America, Mexico, and in a few places in North America. (p. 4) At the same time, they had the foresight to bring with them plants, domestic animals, and tools with which to start life anew in strange environments. Their courage to explore the unknown world was matched by their practicality and know-how. Moreover, the symbiotic knowledge of the sea, its dangers and resources, as well as their morphogenetic invention of seaworthy vessels that could survive storms, showed a genius for survival.

From 12,000 to 10,000 years ago, the world grew warmer producing great floods. Temperature increases and heavy rainfall raised the levels of lakes and rivers. Consequently, land plants and aquatic life flourished. It is estimated that perhaps 15,000 languages already existed worldwide. (p. 4).

Surely the languages spoken must have reflected the morphogenetic investigation of nature. Even today the most primitive tribes can identify in detail the plants, fauna and geographical features of their ecosphere. This ability shows considerable symbiotic capacity to accumulate knowledge about the environment they lived in. Undoubtedly, our ancestors' keen perception of nature gave them quite an exact knowledge of species. Hence morphogenetic perception and symbiotic apperception helped humankind evolve an intelligence nurtured by our own inborn morphogenesis and symbiosis. Human intelligence became a synchronization of these two processes.

About 8,500 B.C.E., people in prehistoric Palestine built not only large stone walls and towers without mortar but also ditches to defend their farming settlements of beehive shaped huts. (p. 5) This little historical fact is revealing of humankind's further development. The stone construction itself and the settlement defenses display not only practical ingenuity and foresight of future needs, such as food and self protection. The beehive shaped huts represent more durable shelters. More interesting is the obvious inference that the design was learned from insects, which, in turn, shows how human ingenuity learned design from nature.

From 8,000 to 5,000 years ago, food cultivation and herders were found in today's Near East, Turkey, and the Balkan Peninsula. (p . 6) Clearly such skills required symbiotic knowledge of soil, climate and duration of sunlight as well as the domestication of animals.

From 8,000 to 4,000 years ago, many fishing communities lined the rivers and lakes in upper Africa and in West Africa. (p. 6) Here again we have an example of how a simple survival skill enhanced human intelligence. A bit of thought about the act of fishing is in order. If the fisherman is in need of great patience and a certain knowledge of the habits of water creatures, fishing itself involves hooking the hidden and the evasive. Over time did this ancient skill teach him patience and self-discipline? Such traits are the rudimentary basis of human conscience.

From 7,600 to about 6,000 years ago, people in permanent villages had learned to grow staple crops like barley, beans, lentils, peas, and wheat. (p. 6) Obviously this productive use of nature was based on the observed characteristics of plants. The elementary symbiotic knowledge was based on humankind's observation of the metamorphosis of seed or bean into plant. More importantly, they learned both to nurture it and symbiotically to cooperate with its hidden, generative power.

There is archeological evidence that the people of the time kept count of trade items, harvests, measures of grain, numbers of animals by using small clay cones, disks, and spheres as a kind of counting system.

Of particular interest in this historical fact is that the association between an object representing a number of animals or quantities of grain is a form of symbiotic knowledge. An equation between mind and outer reality or between a material thing and its abstract representation seems a kind of curious, mental mutualism characteristic of all knowledge we possess. Without one another, neither exists.

From about 7,500 to 7,000 years ago, humans everywhere learned to navigate and manage rivers and lakes as well as dig ditches for purposes of irrigation. (p. 6) To be sure, water itself meant life. Yet using rivers and lakes meant harnessing a power of nature for man's own purposes. Of course, irrigation demonstrated foresight as to future needs, controlled by man's ingenuity. However, the foresight was probably based on hindsight in remembrance of the devastation that torrential rains or rampaging rivers could cause. In a superstitious sense, the spirit in the water needed to be tamed and at times worshiped and placated.

From 6,500 to about 4,500 B.C.E., the potter's wheel was used in Asia Minor. Grain farmers and agricultural communities were widespread in the Yellow River Valley (China) , the Indus Valley (Southeast Asia), Gulf of Tonkin (South China Sea),Nile Vall ey (Egypt) and Mesopotamia (Tigris Euphrates Valley).

About 6,000 years ago the Metal Age began. Metallurgists made copper and gold beads, spears and arrow tips, hooks and trinkets. Metallurgy was practiced throughout Armenia, the Balkans, Turkey, Mesopotamia, Sinai, Western Asia, and Persia. (p. 7)

It is important to note that metallurgy is a process involving a number of steps: smelting (melting the ore), refining and pouring it into dies so as to create all sorts of shapes and forms. To purify the raw ore is to extract the pure metal. Obviously, the entire process involves a morphogenetic transformation executing a methodical symbiotic plan.

Around 4,236 B.C.E. the earliest Egyptian calendar was made. (p. 9) Throughout history the creation of calendars demonstrated the importance of time to humanity. Calendars record significant religious, historical or cultural dates in order to commemorate them. Calendars make clear the cycles of time in human life. So although morphogenetic growth and decline characterize all animate nature, yet the cyclical remembrance of memorable or sacred past events manifests a new form of symbiotic intelligence and knowledge, The calendar serves to maintain the identity and integrity of a group or people.

The Anglican archbishop of Armach in Ireland, James Ussher (1581–1656), declared the year 4,004 B.C. to be the time the world was created by God according to Biblical sources. While such a declaration may be futile to prove time truly began with the inception of one's faith or religion, the assertion was an indication of a growing consciousness of the significance of time. As generations passed, the calendar likely helped provide humankind with an awareness that a lifetime was not to be wasted. Indeed, through the commemoration of past events, thoughtful individuals may have realized that dedication to some life philosophy could give one's life a greater meaning.

For the Hebrews, according to some Rabbis, the year 3761 B.C.E. was the year of Creation, precisely upon October seventh.

Here a calendar date is used to indicate the start of a serious, unified faith. For the faithful the date announces the spiritual birth of their people. It meant the individual must strive for a higher destiny than merely existing and dying.

With calendars based on heroic or sacred events, there emerged a sense of destiny, and the calendar served the purpose to remind successive generations of the "everlasting" truth of symbolic, historical events which shaped their humanity. As the cyclical disappearance and reappearance of the sun, moon, stars and constellations, the calendar is an enduring record of what humanized us.[3]

The year 3641 was the year of Creation for the Mayas, according to some experts. An example of humanity's universal intelligence, that notation of time in Mesoamerica is again evidence that mankind was growing aware that time had a human meaning. The commemorated passing of time reminded them of the dead and the passing of one's own life. At the least, the realization made some minds ponder the meaning of life and its purpose.

During the period 3,500 to 3,200 B.C., the Sumerians marked time with a lunar calendar. (p. 11) It may be a coincidence but the calendar noted not only the cycle of the moon but also the twenty-eight day period of ovulation of the human female. For women this concurrence must have established some kind of natural or mystical connection between the moon and womankind.

The remarkable return of constellations, the signs of the zodiac,[4] were more than imaginary projections. To the ancients, they guided the destinies of humankind. Whole cultures believed in a supernatural conjunction between heaven and earth.

From 3100 B.C.E. to 4100 A.D. tax assessors used geometry and arithmetic to calculate taxes. The first month calendar was created. Moreover, the hieroglyphic, pictorial language invented showed morphogenetic ingenuity, serving a practical purpose. The calendar guided the agricultural life of farmers.

About 1500 B.C., Semitic peoples in Palestine created the first alphabet by simplifying the Mesopotamian cuneiform characters to only 30 phonetic signs. This alphabet was quickly adopted by Syrians and Phoenicians.

As a process, simplification reduced needless complexity into a more efficient and economic means of language writing and overall communication. The ready adoption of the alphabet also showed the eagerness of people to learn what was a valuable new skill. The adoption of writing served the symbiotic purpose of uniting efforts to achieve a common purpose.

From 3,500 to 3,000 B.C., the Sumerians built ziggurats. The flat-topped construction at Ur makes clear they understood the use of arches, columns, domes, and vaults in their architecture and building. (p. 11)

Not only do these skills show the Sumerians had a strong streak of morphogenetic originality but also they could visualize how shapes and forms fit together symbiotically. (Such structuring is quite similar to the shaping and designing skills of the cells in our own bodies.)

The Sumerian number system was also based on 12, 60, and 360, which shows a sense of proportion and the regular interconnections of numbers. (p. 11) Ingenuity in the use of numbers is one thing, but sensing their symbiotic proportion, asymmetry ($5 \times 12 = 60$) and symmetry ($6 \times 60 = 360$) is another.

Craftspeople and artisans started making linen from the stalk of the flax plant. This demonstrated ingenuity based on symbiotic knowledge of the plant.

From 2,300 to about 1,500 B.C., Stonehenge was one of thousands of megaliths built in western Europe. (p. 16) Archeologists have long speculated as to their purpose. Possibly they were a sign of humble submission to the mystery of existence. More scientifically speaking, perhaps those laid out in straight lines across vast regions traced magnetic lines of force in the earth. Perchance, the circle imaged the cycle of time and the circle of human life itself. Conceivably, Stonehenge was the symbol of the Supreme Being's omnipresence on earth and in heaven. Perhaps the monumental stones were a form of worship to the eternal power in the sky.

For early man, life was transient and ephemeral. The fertility and abundance of spring and summer were followed by the withering and dying of fall and winter. The next spring the cycle renewed itself. Darkness was followed by light, death by life. A man's earthly destiny seemed circumscribed by the eternal return of sun, moon and stars.

Today we wonder if the morphogenetic/symbiotic life force in humankind, in its own way, is eternal, capable of generating life anywhere in the cosmos where abiotic conditions allow it.

ಸಿಂ

From 2,000 B.C. a diet of bean, squash and corn was common in Mesoamerica. (p. 19) Growing crops successfully took ingenuity and experimentation mainly by women to discover which plants were "weeds" and which plants were edible with cooking, not poisonous but nutritious. Usually men were so involved in adventure, exploits, exploration and defense of the group or community, they hardly realized the variety of creativity women brought to bear in domestic life. Because of women's constancy and purpose to keep a family together, they had to develop their own practical creativity. Moreover, their understanding and sympathy enabled them to keep together the family long enough to ensure the safety and education of their children to young adulthood.

ꙮ

The Minoans, like many ancient peoples, worshiped a "mother goddess" and made painted pottery. Her worship implies she was desirable for her fertility and creative power, and she was respected for her maternal sense of responsibility in the care of her children. These inborn qualities made her instinctively attractive to the male.

Moreover, the pottery was often shaped like a woman's hips, for a vase was made to hold grain, seed, and water as life-giving sustenance. So pottery represented symbolically the abundance and promise of future generations to come.

ꙮ

From 1,900 to 1,500 B.C.E. was the pinnacle of Minoan civilization on Crete. The Minoans had running water and sewers. They decorated their walls with frescos. Their merchants exported wine, olive oil, pottery, gems, textiles, tools, weapons, and luxury craft wares; they imported metals and foods. Their main deity was the Earth Goddess, Rhea, the mother of Zeus. (p. 20)

The evidence of morphogenetic ingenuity, innovation and invention is obvious in the drainage system, the frescos, the textiles, tools and weapons, . However, the extensive importation and exportation of goods and food clearly show how far symbiotic exchange had already evolved some 3,000 to 3,500 years ago.

The Earth Goddess was worshiped because she was the source of all life and present at the birth of all living things. In her were illustrated three qualities: (1) she was an ancient symbol of nature's fertility; (2) she symbolized the symbiotic oneness of nature; and (3) she represented the infinite patience and power of maternal love.

ꙮ

About 1,800 B.C.E. the Sumerians and Babylonians knew how to calculate square roots and cube roots as well as do some geometry and algebra. (p. 21) Examples: The square root of 9 is 3; the cube root of 27 is 3, which equals 3 x 3 x 3. They also converted their cuneiform writing into ideograms.

Not only is such mathematical reasoning ingenious in the manipulation of numbers. Such experimentation is clearly a form of mental morphogenesis. Moreover, transforming cuneiform writing or pictographs into ideograms is evidence of symbiotic simplification of design, as we often see in nature (e.g.,the common leaf). Yet, in addition, it shows how symbiotic reasoning can use two seemingly opposite processes effectively: (1) from specific to the general (from intricate complexity to greater simplicity and efficiency or (2) from the general to the specific (as from a vague notion to a more exact idea). Consequently, shapes become ideas that eventually become meanings. Or as in deductive logic, a generalization, (the major premise of a syllogism) tested by a minor premise (an example), can be made an exact statement of fact.

Laws, Rights, and Civilization

In addition to the manifestation of ingenuity, humankind have throughout history been preoccupied with establishing ways to govern human conduct and societies. Nineteen hundred years before the life of Christ, the king of Babylon, Hammurabi, had some 282 laws codified in stone. Based on the Sumerian idea of justice and the legal ethics of Assyrian, Chaldean and Hebrews (p. .21), the code specified the law of reciprocity—value for value, loss for loss. In short, it was the famous law of "a tooth for a tooth, an eye for an eye."

It is interesting to note that the code is a severe form of symbiotic thinking with its emphasis on law, order and justice. In Hammurabi's time, the tough, legalistic side of justice was stressed. The code was needed to regulate the relations of humankind and communities. It was an example of Conscience as our ancestors most clearly understood it. This ancient system of law and order reminds us that there should be justice and punishment where it is deserved, but punishment in accordance with and proportionate to the transgression, "sin"or crime.

In the fifth century before Christ, ancient Rome established the Law of Twelve Tables, which was a code of civic liberties and duties. Centuries later the Magna Carta of 1215 was a charter of liberties wrested by English barons from King John. It guaranteed them their fundamental rights and privileges.

Perhaps a curious example of law was John Calvin's Institutes of the Christian Religion (1536). The main idea was that congregations should elect their own ministers. Given the history of the priesthood, this was a rather startling idea. It meant that the faithful had the right to be led by the most intelligent and most knowledgeable among them. They could elect their own mental and moral mentors.

Of course the term law had other, extended applications. A shift in meaning comes about when thinkers began to speak of scientific laws. For instance, when Johannes Kepler (1571–1630) published *Mysterium Cosmographicum*, he spoke as if God were an astronomer who had provided the laws of planetary motion. (p. 153) To be sure, the regularity of planetary cycles showed they were governed by laws. This understanding may have led to the insight that there might be laws that should govern human life above and beyond legal systems.

Kepler's work was followed by Isaac Newton (1642–1727), who explained the law of motion and gravitation. (p. 166) Moreover, his invention of the telescope and calculus provided him with tools to extend his knowledge. Thus he used ingenuity to discover greater knowledge.

At the same time the Dutch naturalist Anton van Leeuwenhoek (1632–1723) was exploring microscopic life. (p 159) Here again, human ingenuity had invented an instrument to increase our knowledge of the world. In the following centuries, the significance of the microscopic universe revolutionized our vision of nature and life itself. The twentieth century demonstrated the omnipresence of universal bioprocesses inherent in all forms of life. Indeed, these bioprocesses not only describe our cellular nature and genetic history but also explain evolution as a dialectic between the morphogenesis and symbiosis intrinsic to all that lives.

Let us now examine the manifestation of civilization in human history. It offers us a kind of eco-cultural understanding of humanity's relationship to its world.

In ancient Egypt, a century before Christ, there arose the cult of the pharoah. Also worshiped were a number of important gods such as the Re, the sun god; Hathor, a fertility god; Osiris, the god of death and the afterlife; and Thoth, the god of the sciences, writing, and the law. (p. 25)

It is noteworthy that each god served a distinct human need since the individual could appeal directly to a special god for help. The symbiotic one-on-one relationship seemed to assure the supplicant of the god's concern and aid. Moreover, even in death there was hope, for if one worshiped the god properly, he would control what happened to the deceased after death. The god would decide the petitioner's reward of life after death.

Yet beyond this down-to-earth religion, other cultural facts were important. In a land where the desert represented barrenness and death, the people must have felt gratitude for the fertility and creativity of the land flooded by the Nile River. In addition, they must have been conscious of the perfection of forms in the green flora of their fields and gardens. This knowledge of natural forms is seen in the art, sculpture, and architecture of their temples. Obviously their craftsmen emulated nature's precision and grace.

Similarly their worship of the god of writing and science (Thoth) makes evident the importance the ancient Egyptians gave to designing their hieroglyphics and to their investigations of nature. Put another way, Egyptian civilization may be characterized by a morphogenetic drive to gather knowledge and by the symbiotic motivation to complete and perfect whatever skill they used. History tells us they rewarded well their veterinarians and physicians. Furthermore, as proof of their intelligent symbiotic reciprocity, they established centers where astronomers, geologists, surveyors, and scribes gathered to exchange knowledge.

Ancient Roman civilization spanned more than 2,000 years of existence. It passed through stages of kingdom, republic and empire. Each period of history had its own philosophy of government. To be sure, the superbly engineered public works (monuments, bridges, aqueducts and road networks) expressed an immense organizational ability. Their ingenious application of architectural techniques showed morphogenetic imagination and the symbiotic integration of solid structures that lasted well over a thousand years.

On the other side of the world, Middle America (300 B.C.E. to 900 A.D.) the Mayan people developed their own civilization. The pyramids, stelae and temples served ceremonial purposes, both public and religious in nature. (p. 68) As the ziggurats of Mesopotamia and the pyramids of the Egyptians, the Maya structures served the symbiotic purpose of unifying the populace and the leaders by a spiritual bond. Through the construction of such religious monuments, the people were taught there was a higher power in nature, which must be obeyed and served.

The Far East supplies our last example of civilization. For thousands of years China has been a single civilization. With one basic culture and one writing system and the majority of the population speaking one language, Mandarin, China exemplifies one of the most enduring civilizations on earth. Such symbiotic unity over time indicates a pervasive consolidation of culture. Although the Chinese invented many useful things and created a very rich art, literature, and philosophy of its own, their symbiotic, conservative tradition largely dominated its morphogenetic inventiveness for many centuries.

We can now sum up what established civilizations demonstrate in terms of nature's two ubiquitous processes. Each civilization demonstrated an extensive use of morphogenetic ingenuity, invention and creativity as well as symbiotic coordination and subordination in their social organization. Evinced was a remarkable integration of human effort to pursue a common cultural objective as well as maintain the safety and security, the health and well-being of the majority of the population. Through individual self-sacrifice, the good will derived from sharing a language, the people were motivated by a mutual purpose to preserve their cultural identity. Thus did civilizations endure.

In sum, we have presented historical evidence that morphogenesis and symbiosis pervade human ingenuity, law and civilization. Furthermore, this testimony illustrates the fact that these universal processes have been at work in virtually every phase and stage of human evolution.

Conscience in Past Life Philosophies

Although Lovejoy's, *The Great Chain of Being,* gave a convincing presentation of the interaction between the archetypal processes being and becoming, it would be hard to prove these concepts effected the evolution of human thought. To be sure, over the centuries, refinements and extensions of their meaning did emerge as succeeding intellectual generations re-examined these nuclear ideas. But whether they influenced the lives of common humanity is all together another matter.

The same objection regarding influence cannot be made as to the various religions embraced across the world. Indeed, although some religions offered a cultivated way of understanding nature, others concentrated more directly on the ethical education of human nature. As such, they influenced humanity over thousands of years. They have taught us the moral consequences of our conduct. They have given us a deeper insight into the meaning of life.

However, let it at once be understood that we do not expect to find any evolution of morality or conscience through successive world religions. So we will not proceed in any chronological order. To do so would be to mislead the reader that mankind have evolved morally over the past five millennia, which we have not. Rather, at diverse times in human history, quite startling moral philosophies have been proposed, some of which still have a deserved impact on human conscience today.

Hence the purpose of the last section of this chapter is to illustrate how belief, religion, and concept have given rise to life philosophies, past and present. Moreover, culture heroes, founders of religions, and sages have also left indelible memories in the mind of mankind and womankind, which to this day inspire their descendants to live more worthwhile lives.

The 5,000 year history of China provides us with a view of existence and of human life still viable and vital in our day. Some 4,800 years ago, the Chinese emperor FuHsi conceived of nature and life as composed of two eternal principles: yang and yin. The intimate interaction of these two created all that came to be in the world. Yang was the active male principle whereas yin was the passive female principle. As polar forces in the universe, together they brought about order, balance and change. They affected each other cyclically, at one time dominant and the other quiescent. By their reciprocal reactions, they maintained and yet changed nature and life.

The reader will recognize the kinship between yang and yin and morphogenesis and symbiosis. It would seem that the discoveries of modern biology, ecology and microbiology corroborate the ancient intuition. Taoism seems a philosophical hypothesis which five millennia later modern scientific research proved basically true.

In China, about 2,200 years later, the great Laotzu (sixth century B.C.E) is said to have created the Tao Te Ching (Way of Power) to explain the central principle of the universe. It also taught humankind the natural way to live. All living things were animated by *chi*, which is the vital cosmic energy we need to nurture and maintain physical, spiritual and mental balance. The mystical aspects of this philosophy are explored in the *I Ching* (Book of Changes) used today to make divinations.

In the same period, the Chinese philosopher Confucius defined the individual's place in society, in the world and in heaven. Tradition and order were to be respected in order to help maintain the equilibrium of human existence and the universe. The individual was advised to practice *Jen* or benevolence to others. Virtues such as love, integrity, loyalty, and altruism were to be achieved by following the "Middle Way" of inner harmony and balance. The Middle Way is guided by *li*, which primarily emphasized a respectful attitude in following customs and in performing rituals. Above all, the individual needed generosity of soul, sincerity, and seriousness of purpose. Of particular note is Mencius's later development of Confucianism. Mencius taught that the human being should seek moral perfection through study, discipline, and the cultivation of one's inborn energies.

With some thought it becomes evident that Chinese ethical philosophy believed in some mystical correspondence between the individual and nature. It may have derived from an earlier form of animism, but the essential belief was that the human being should cooperate with and emulate the invisible processes that made life itself possible. Beyond the honest recognition of sexuality, the male-female relationship seemed to personify all nature. In fact, this symbiosis of distinct beings was vital to the continuation of life itself. While every man and woman had a definite role to play in the drama of existence, each had to come to terms as well with their innermost being. If the common man and woman was barely aware of this mystical source of meaning in their lives, yet his instinct and her intuition found fulfillment and peace in their life-giving conjunction. Such was their mutual need.

❧❧❧

More familiar to the Western reader is the name of Moses, the Ten Commandments, and Judaism. Well known are such Commandments as "Honor your Father and Mother," "Do not steal," "Do not give false testimony," and the like.

The religious and ethical system of the Jewish people, called Judaism gave them a common identity, genealogy, and history. The divine laws that are the foundation of Jewish faith are found in the Hebrew Bible (*Tanakh*), known to Christians as the Old Testament. The Pentateuch, or the five Books of Moses, and the Torah describe the world's creation. On the other hand, the Talmud reviews Jewish history and legend and contains writings on moral, legal and historical topics. Based on covenants between God and the patriarchs, to study these writings and commentaries is to follow the spiritual path to knowledge of God.

Obviously, the Ten Commandments and Judaism stress the importance of maintaining sensible, honest, decent, respectful, and moral relations with one another. At the same time, the commandments are "higher" than laws because the individual must obey and enforce these injunctions in his own heart and conscience. Represented is a symbiotic conscience emphasizing self-discipline, strictness of obedience, the need for moral integrity and justice. This life philosophy became a declaration of communal faith and relgious solidarity for the Jews. In other words, strict obedience to the Word of God meant obedience to one's conscience.

❧❧❧

Also familiar to the reader is the influence of the ancient Greeks on human thought, history and the philosophy of life. When the famous *Iliad* and *Odyssey* were written about a thousand years before Christ, the warrior code of life expressed their sense of honor and loyalty. On the way home by sea, the epic hero of the Trojan War, Odysseus/Ulysses survived all manner of adventures, temptations, and life-threatening dangers through the use of his cunning and resourcefulness. Though he had been gone many years, his wife Penelope held off brazen suitors, and thus throughout Greece, she became the model of the dutiful wife, loyal in love and firm of conscience.

Yet ancient Greece also gave us the philosopher Socrates (470–399 B.C.E.), who developed a method of probing for definitive truths aimed at countering the unscrupulous arguments of the sophists and the skeptics' denial that truth was knowable. Investigating concepts via skilled, leading questions, Socrates' dialogues led to meaningful definitions. Thus by his use of irony, Socrates could lead his interlocutors to understand the need for excellence, or *arete*, by which he meant there was a best course of action to be taken. Unforgettable are Socrates' words "The unexamined life is not worth living."

Thus a single truth can endure millennia and test the strength and integrity of every man and woman. The injunction tells us we alone can discover who and what we are. It tells us that only our own conscience can guide us to live a life intelligently. Only through self-honesty and self-knowledge can we master whatever fate brings us.

Hence, so many centuries ago, we have in Socrates a single, strong mind able to shape a civilization. His dialogues were of the nature of a symbiotic encounter, the exchange of original points of view with a mutual purpose: to achieve a consensus, a definition of terms, a synthesis of viewpoints. Yet his stance stood for something more. His insistence on re-examining with care our assumptions and our most basic knowledge taught us to investigate nature more exactly and to reexamine what hitherto we had thought was human nature.

※

The ancient Greek philosopher Plato (428–347 B.C.) believed that civil strife and moral corruption would never cease without understanding the purpose of life. His *Republic* examines what is meant by the good life. The key means to that life and to a stable society was education.

His spokesman Socrates often used analogy to argue effectively. At one point in *The Republic* he makes clear that injustice leads to human disunity and to civic chaos. Moreover, justice is needed as much as are true knowledge and good health.

Plato is concerned with creating the Ideal State. He believed the individual was happiest when he worked at what he was best suited to do. Plato urged the moral education of children, and he submitted that "golden parents" have "golden children." Moreover, the better the education of the people, the fewer laws will be needed to run the state.

Four Cardinal Virtues should govern humankind: discipline, courage, wisdom and justice. Discipline meant not only temperance but also "self-control" and "self-mastery." Courage meant strength of character and the readiness to sacrifice oneself in battle, if need be. Wisdom meant the ability to judge right from wrong, good from bad. Most interesting was justice. It made possible the other three virtues. A just man's reason is in control of his emotions and desires. A just man or woman has an orderly mind or soul, yet each part of the body and mind must allow for fulfillment. Plato stresses that women are born with aptitudes and abilities as men, and women deserve the same quality education as men. The aim, for both, was an integrated personality. The philosophical ideal was that each individual be educated in the cardinal virtues.

From this brief sketch of Plato's *Republic*, we see clearly that western civilization was positively influenced by the moral life philosophy he proposed. His reasonable and noble outlook offered every man and woman a plan for achieving an intelligent destiny.

❧☙

Ancient Greece demonstrated a remarkable development of the philosophical conscience. The philosopher Aristotle (384–322 B.C. E.) studied nature as would a modern biologist or zoologist. Next to Plato, he became the other philosopher to dominate Western thought for the next two thousand years.

His investigation of nature was inductive. He maintained that observed facts must precede theory. Nevertheless, he formulated the special logic called the syllogism, which used a three-step method to come to certainty of conclusion. He started from a generalization (the primary premise), applied it to a specific case or example (the secondary premise) to reach a definitive conclusion. An illustration of a syllogism would be the following: (1) "All men are mortal." (2) "Socrates is a man." (3) "Socrates is mortal."

On the other hand, Plato had considered the physical world an illusion by contrast to the eternal ideas and forms conceived by a philosopher. Yet obviously, plants and animals in the real world continually changed and suggested another "eternal" truth.

Aristotle overcame Plato's dualism by the concept of potentiality (e.g., a pine cone has the potential to become a tree) and entelechy, the actualization of that potential (e.g., the actual growth of the pine tree itself). Aristotle's solution to Platonic dualism was replicated in the history of ideas as *being* and *becoming*, which theme preoccupied European thinkers for the next twenty-two centuries. Finally, Aristotle's insight into the invisible potential and visible entelechy of all life prefigured the twentieth century discovery that morphogenesis created and symbiosis designed everything that lives.

※

The ancient Greek Epicurus (341–270 B.C.E.) developed the philosophy called Epicureanism, based on the doctrine that pleasure and happiness are the main goal in life. At the same time he considered emotional calm to be the primary virtue in the pursuit of intelligent pleasures. However, rather than pursue sensual satisfactions, he advocated temperance of body and serenity of mind. He opened his academy to freemen, women and slaves alike.

※

Another ancient Greek, Zeno of Cyprus (335–263 B.C.) introduced Stoicism. This life philosophy believed a wise man should remain free from excesses and passions, be unmoved by joy or grief, and obey natural law. The stoic should firmly control his response to pain and distress. Reason and good conduct were essential. The asceticism of the stoics earned them the reputation of courage, self-control and wisdom. Stoicism not only became widespread through the Greco-Roman, Hellenic world. Over the centuries it influenced men and women of principle as well as defenders of human rights.

※

Outside the GrecoRoman world there were other forces at work which also influenced the history of humankind. These forces were concentrated in various religions, each of which provided a particular understanding of existence. Eventually, they were to influence life philosophies even of modern man.

For instance, 1000 years before Christ, there arose Zoroastrianism in Persia, which religion marked the history of Judaism, Christianity, and Islam. Its founder Zarathustra rejected the pantheistic worship of his time to initiate a vision of existence as a cosmic struggle between opposing Powers. On one side, the true god Ahura Mazda (Lord of Wisdom), who created the world and its creatures, represented justice and truth. With his warrior angels (*ahuras*), he opposed Ahiman (Spirit of Evil), who brought evil and destruction into the world. Ahriman was backed by his army of demons. Humanity must choose between them.

Zarathustra foresaw the end of the world in a holocaust, from which eventually good could be born. For the individual to merit an afterlife of salvation, rather than damnation, meant he or she must act for the good in this life.

This existential and moral conflict should sound familiar. It is both implicit and explicit in the life philosophies born of Judaism, Christianity and Islam.

※

Christianity is based on the belief that Jesus is the Christ or Messiah. To Christians he is the son of God in human form. He was crucified on the Cross, sacrificed to atone for human sins and to ensure salvation for the faithful. His resurrection after crucifixion symbolized the promise of an after life and an everlasting spirit in God's presence.

The teaching of Jesus included practicing charity, mercy,and justice. Especially important was the expression of brotherly and sisterly love to one another. From his Sermon on the Mount,two admonitions emerge. The first was "Do not judge, or you too will be judged." Here he advises the intolerant that they should try to better understand others. Otherwise, they in turn will be judged with the same severity. The second admonition was do to others what you would have them do to you, for this sums up the "Law and the Prophets." Hence reciprocity in kind is advised. In other words, the life philosophy of Christians is based on mutual tolerance, empathic understanding and moral charity.

Evidence for the view of existence as a sphere where forces of good and evil are locked in mortal combat comes from Christian history. For eighteen centuries after the life of Christ, religious literature, art and architecture portrayed the titanic struggle between the power of God and Satan, the angels from heaven and the devils from hell. Throughout European Christendom, the theme of *psychomachia* in painting and sculpture, sermon and religious instruction personified "resplendent angels" as virtuous and the "vices" as sordid, hideous creatures out of hell.

Thus existence seemed a struggle for the conscience of humankind. Its outcome could not only condemn you in this life but damn you in the life hereafter.

※

Islam is the religious faith of Muslims. The religion is extensively observed by one billion followers worldwide. Islam maintains that Muhammad is the last of the holy prophets mentioned in the Old Testament including Adam,

Abraham, Moses and Jesus. To be a Muslim we must surrender to the will of God and demonstrate total devotion to Allah, the one and only God.

Muhammad's authority came from God directly as transcribed by the Qur'an (Koran). This scripture of Islam is the basis for Muslim custom and law. The Five Pillars of the faith prescribe the practices to be followed: five prayers a day while facing Mecca; charity to the needy; fasting during Ramadam in remembrance of the impoverished and hungry; and a pilgrimage (Hadj) once in a lifetime.

Central to their belief is that God is just and merciful as the Creator of the entire universe. For many Muslims the world contains *jinn* or demons whose temptations the individual must resist and overcome. Moreover, it does have a warrior credo in the *jihad*, the holy war which Muslims may wage as a religious duty. Nevertheless, the core of their life philosophy in the Qur'an is forgiveness. Thus a duality of conscience seems to rule.

There is an esoteric branch of Islam in the mystical practice of Sufism. Convinced that God is the ultimate truth of existence, the Sufi exercises self-denial and intensely devotes himself to evading life's delusions and suffering. The Sufi aim is to attain ultimate enlightenment through progressive stages of repentance and renunciation. Note in this philosophy that the source of pain and suffering in life is to be overcome by asceticism, self-sacrifice, and a turning away from life itself to nurture the soul. Thus the Muslim mystic's conscience.

※※※

We must not neglect two major Oriental religions. For more than five millennia, Buddhism and Hinduism have influenced the life philosophy of billions. Buddha, "the Enlightened One" (563–483 B.C.E.), was the religious philosopher who founded Buddhism. One day he was converted from a superficial life of wealth and luxury when he encountered an old man, a sick person, a dead body and a mendicant monk. These "omens" warned him to renounce his former life so as to become an itinerant beggar. He undertook to discipline his body and mind. Discovering rigorous asceticism did not bring the spiritual enlightenment he sought: through contemplation he found perfect Enlightenment or Buddhahood. When he was ready, he introduced the principle of Dharma to a few disciples as the way to enlightenment.

Dharma includes three key ideas: (1) the individual should fulfill his duties by observing custom and obeying the law; (2) the basic principle or divine law governing the cosmos and individual is one, therefore, it should be obeyed; (3) one's duties should conform to one's true nature. By following the Dharma one attains *moksha* or spiritual liberation from the cause of pain and suffering in the world.

Important to understanding both Buddhaism and Hinduism is the need to grasp the meaning of karma. Both believed that our actions in life generate moral consequences and perpetuate the transmigration of one's soul into another life of suffering, that is, a new cycle of birth, death and rebirth. To break free from this endless self-perpetuation, one must rid onself of the bad karma still with us from previous lives. Liberation comes only by submitting to a life of austerity.

Put another way, Buddhism teaches that suffering is inherent in life, yet one can free oneself from it by mental and moral self-purification. Ignorance and desire cause our suffering, but by "right" behavior and action and by moderate speech and contemplation, we can overcome desire. Enlightenment enables one to slough off one's karma. Through Enlightenment, one reaches a spiritual state free from attachment, delusion and suffering. In other words, one can attain nirvana, the ultimate beatitude beyond suffering, achieved through the extinction of desire and individual consciousness.

As noble and worthy as the enlightened path may be, it seems too preoccupied with one's own freedom from the karmic cycle and too concerned with one's own "salvation." Later, the Mahayana branch of Buddhism emphasized the need for Bodhisattuas in this world. Such beings took on suffering and sacrifice for the good of others in order to save them.

In the context of *The Creative Conscience*, such altruism represents the highest form of symbiotic empathy and caring for others. A degree of such selflessness is an antidote to the selfishness in the world, as long as the selfish do not simply exploit the selfless. Expressed in less spiritual terms within the reach of Everyman and Everywoman, selflessness is the chief commitment of heroes in every walk of life and is the secret dedication many men and women give to their profession and calling. If nirvana is not in reach for ordinary mortals, the essential meaning of our lives can be realized through a personal dedication to humanity.

༄༅༅

More than 4,000 years old, Hinduism has no fixed doctrine, no single authoritative scripture, or official "organization," yet today it has more than 500 million adherents. Brahma is their creator-god, which is personified by brahman, the priesthood and the highest caste among the Hindu. As the ultimate ground of being, Brahma is the source and essence of all life.

For the devotee, the goal of this life is the union of one's soul (atman) with brahma, the supreme universal self. The goal, (*moksha*), is to find release from samsara (the perpetual wandering of the soul through the cycle of birth, death and rebirth incarnate as plant, animal or human). Each incarnation is determined by your karma or moral record from past lives. For better or worse, your previous incarnations decide what your next life will be.

One can overcome karma by obeying Dharma, the moral law through each stage of life. In the Bhagavad Gita (Song of the Lord), the Lord Krishna (a hero deified as the preserver god Vishnu) provides the most significant lesson of Hindu belief. Krishna encourages all who seek the Way that there are many acceptable paths to salvation, but each person must seek what is appropriate to his soul.

Figuratively speaking, Brahma could be the source of the creative morphogenetic power in nature and the integrative, symbiotic power that organized life on earth into its myriad forms. Again, metaphorically speaking, Vishnu might express earthly fertility and the symbiotic unity that only love can bring. The eternal togetherness of nature's powers is revealed in the earth's great cycles, bringing ever new life into being.

The Enlightenment

The infinitive "to enlighten" means to teach knowledge which will bring the student or adept to comprehensive understanding. To be enlightened is to understand fully the complexity, intricacy and essence of a problem, principle or theory.

In Europe and America, the term Enlightenment describes the rejection of assumptions and ideas esteemed in the past. A new reliance on human intelligence and reason began to supercede blind faith in religious truth. Reason became the new source of knowledge superior to any sense perception. Hence the Age of the Enlightenment (eighteenth century) no longer trusted traditional authority, emphasized the secular, stressed human dignity, and believed reason would illuminate humankind's future.

In contrast to the tradition of Eastern religions (Buddhism and Hinduism), which neglected the pursuit of earthly knowledge and gave up any hope of finding happiness or peace in this life, the West focused its intelligence and energies on the development of reason and realizing the hope that concerted effort could improve the human condition. As embodiment of this rational faith, the *Encyclopedia* (1751–72) edited by Denis Diderot,sought to define and describe all branches of human knowledge, crafts, trades, and other practical skills.

Deism

Originating in the seventeenth and eighteenth century England, Deism was an intellectual movement that basically rejected the canon and dogma of traditional Christianity. It disbelieved that the Creator interfered with the laws of the universe He created. Deists disputed revelation and miracles and rejected any supernatural explanations of earthly events. Consequently, they were skeptical that God interceded in human events.

Thus they denied divine providence, which meant God did not decide human destiny. Yet, curiously, Deists believed that the creator-god rewarded the virtuous and punished the wicked in an afterlife.

On the positive side, they considered tolerance was a virtue, that man was in control of his own destiny, and that reason should govern human relations. Yet Deism apparently did believe that some universal law of nature guided human nature to develop man's moral conscience. Hence Deism may have been inspired by the Logos, the natural law of the ancient Stoics; the *jus naturale* or the natural law of the ancient Romans; the seventeenth century concept of international law; and John Locke's natural law (derived from nature) which bound human society to ensure all individuals their natural rights.

It should be obvious that Deism and the philosophy of *The Creative Conscience* share certain affinities, especially in the belief that nature provides a source of knowledge to guide humankind to a sound life philosophy.

However, in the nineteenth century, the trust in the benign purpose of nature was abruptly overturned by Darwin's *Origin of Species* which in the popular imagination seemed to portray the fact that nature was governed by a malign purpose. Nature now appeared predatory, a struggle to survive, a food chain of life-and-death, a place where the strong, ruthless and vicious conquer. Darwin's "scientific" perspective seemed to demonstrate a total absence of moral conscience. There was no justice in nature.

Conclusion

In the previous chapter we perceived how divergence and convergence influenced human relationships. We saw how languages, ethnic groups, religions, and entire peoples diverged over time, losing contact with their proto-languages and the customs and traditions they once shared in the past. By contrast, we came to recognize how humankind's search for knowledge, their shared survival skills, their aptitudes and talents for creativity not only enabled groups to survive but also to live in peace and harmony together.

The history of religions manifests a similar divergence and convergence. Our earliest religions were polytheistic with spirits, demons and gods everywhere in nature. To be sure, this superstition may have begun during our prehistoric stage of survival by the camouflage detection we used in order to detect hidden predators in forest and grassland. Hence this survival vigilance may gradually have led to a habit of uneasiness, distrust, and fear of the unknown. Later, humankind became afraid of their own kind.

With Zoroastrianism, Manichaeism, Judaism, Christianity, and Islam, there emerged a predominantly dualistic vision of life. They believed existence to be an internecine war between good and evil. This belief led to the unfortunate consequence of dividing humanity into those who believed as we do and those who did not. We were good; they somehow were evil. Humanity diverged between the civilized and barbarians, allies and enemies, victors and victims. Conflict and war seemed the heritage of humanity.

However, in actual nature there is no such thing as dualism. Yes, there is night and day, darkness and light, but nature is not divided by invisible, opposing powers. Dualism is a non-fact. It is an illusion and a delusion. There are no dualistic powers in existence nor even in our own human psyche. Neither nature nor human nature operates or functions dualistically. If anything, our mind processes vivid imagery and is a reciprocal process between polarities.

To be sure, if driven to excess by external circumstances or by internal biological needs, there can result distortions of truth, addictions, intense passions, warped understandings of life, submission to superstitions, and war within the individual soul or between mankind. But it is fear, anxiety, desperation, or despair that brings about excesses. No invisible power is driving us to our fate.

In so far as religions are driven by dualistic/antagonistic interpretations of the real world we live in, dualism is an anachronism in the history of human thought. It makes a mockery of religion's intended education of the human conscience, because dualism turns all the world into enemies, instead of teaching us the value of life and the need for an educated conscience to negociate human relations. Nor should we accept fatalistically the definition of life as mainly suffering, sickness and death. To believe that our natural joys and gratifications are to be blotted out for fear of what fate will inevitably bring us is a sacrilege against being born to live.

On the other hand, religions have offered very positive benefits to aid humanity. For instance, the polytheism of ancient Egypt and of Hinduism rendered the individual supplicant a source of hope and comfort by having definite gods to appeal to for help. To a degree, Roman Catholicism with its saints also offers the humblest person a mother or father figure which promises saintly protection and guidance. In this way, people without other material or spiritual resources find shelter and consolation. This arrangement stems from a heritage of compassionate conscience.

In turn, monotheism brought with it an evolved form of conscience. There was a definite advantage to believe totally in God or Allah, Brahma or a Buddha. The belief in monotheism had the consequence of integrating both our mental and spiritual energies so as to nurture a holistic vision of life.

When the single reason for existence is a Diety, humanity becomes aware that life itself must have some paramount purpose. The omnipresence of the Divinity would seem to urge every individual to integrate life's experiences into some final meaning. If we are alive, there must be a reason for it. One's own life must have a meaning. Our concept of God is to assure us it is our responsibility to creatively and conscientiously define our own earthly destiny.

ಸುಧಾ

At this point, let us glance back again to the ancient Greeks. In so far as the Stoics taught us the need to resist pain, sickness and suffering, they encouraged us to survive in this world. Moreover, to survive as a Stoic also enables us to help others find the strength and determination to complete their own destiny. Staying alive, we can be of use to our fellow humanity and help mitigate their miseries and misfortunes.

On the other hand, the Epicurean encouraged us to enjoy not only physical but also intellectual and aesthetic pleasures. Obviously, human creativity is a gift our entire species shares. Thus if Stoicism is a good model for a strong conscience in control of one's self-pity and weaknesses, Epicureanism is a good model for the finer pleasures life offers through being creative. Thus, given the history of humankind, it is apparent that all humankind share a keen pleasure in originality and invention whether of an aesthetic or practical nature. It seems that genuine creation will never be cause for any deeply felt resentment or disagreement. Because in our childhood we imagined stories, enjoyed games, and played with one another innocently, that early experience gave us our sense of what life should be. Perhaps that is why creativity is so close to our hearts.

On the other hand, given the history of the human conscience, we have seen how many forms it has assumed. We find that common denominators activated the development and evolution of conscience.

The first and foremost realization is that true intelligence and conscience are universal throughout humanity. That means no race has "superior" native intelligence or "higher" moral conscience than any other. True, various cultures may evince different stages of creativity and conscience just as individuals may exhibit different levels of creative skills or formal education. Such distinctions simply mean that more "advanced" societies (e.g.,in technology, medicine, science) should aid the less "advanced."

Teacher and student illustrate this realistic relationship. Both may have the same level of intelligence, only the student's is still potential whereas the teacher's is actualized and matured.

Put another way, we need to teach each other what is meaningful and moral in our lives so that we can better understand one another. We need to find out how men and women think about each other, about raising children, about family relations, and about the family of humanity.

We might sum up what our human conscience has taught us by providing a list of sayings.

1. Believe what you will so long as you do not deny me my beliefs.

2. Live your life as you will as long as you do not prevent me from living the life I know to be worthwhile for me and my family.

3. Because at heart we all are children of Nature and one Supreme Being, let us help one another live in peace and practice mutual understanding.

4. Let us not malign or harm others because we believe we are superior to any race or religion.

5. Every life philosophy in the history of the world has contributed directly or indirectly to who we are today.

6. Racial superiority or racial inferiority is a myth born of abject ignorance, repeated only by those who would deny they themselves are of the human race.

7. Believe in others as you believe in yourself. At every dawn and dusk, we should express that faith in humanity.

PART VI

THE INTELLIGENT LIFE

10

A Third Millennium Life Philosophy

Darwin's concept of evolution has been widely used as a figure of speech. Scholars, scientists, journalists, and essay writers have applied the idea of evolution to societies, institutions, art, religion, conduct and morals. As used, the figure of speech indicates an almost universal consciousness of the presence of change and transformation in animate nature and in our own lives.

If the immediate purpose of evolution clearly seems survival to live another day, then we must consider the corollary phenomenon of living a long life. If the prolongation of life is a goal of improved evolution, what would be the purpose of living longer?

For the majority of people, the chief merit of more years would be to continue such hedonistic pleasures as food and sex. However, for some, the main merit of a long, healthy life would be to have time to extend one's aptitudes and perhaps even perfect one's talents. A period of retirement would enable the thoughtful individual to come to terms with the meaning of one's own life. For another, it would be to accumulate insights about life itself or to reconcile oneself to the end. For others, a longer stay would enable them to share their life's experiences with their children or descendants. Finally, maturity and age should bring a measure of wisdom.

In the context of living a long life, the mature individual learns to distinguish between short-term pleasures and long-term gratifications. With maturity and perspective, one gradually gains a sense of control of oneself and over one's own destiny. One learns to convert multiple, superficial (at times intense) pleasures into accomplishments worthy of one's education and intelligence.

To be sure, the theory of evolution stresses that survival depends largely on adroit adjustment to the demands of the environment. However, beyond the obvious benefit of transforming that environment to one's biological needs, the human being has an equally pressing requirement to fulfill. That is to explore one's own personality for its intimate needs and, accordingly, to plan a sensible life that is worth living. Eventually, one becomes aware of a duty to one's intelligence and knowledge to better understand the deeper sense of life itself.

Thus far human nature has been described as activated by morphogenetic curiosity, capacity for self-transformation, and creativity. This process is complemented by the symbiotic instinct to integrate, complete and perfect what one does or makes. Since these two processes interact according to need, they must influence explicitly and implicitly our moral philosophy of life. The purpose of this chapter will be to discuss how morphogenesis and symbiosis effect such a philosophy. But first we must consider related topics.

Since such men as Jean Jacques Rousseau and Mahatma Gandhi resoundly rejected our assumption that civilized man had a "better" life than traditional man, there is reason to explore the distinction. If any, what advantage did early man and woman have over their later "civilized" bretheren? From what we know of traditional societies, the great advantage of living in harmony with nature was that humankind generally felt assured of a degree of protection in obeying the Spirit of Nature. Some humans prayed when asking permission to husband the land. They even prayed for forgiveness when it became necessary to kill any living creature. In such acts of conscience, they demonstrated their sympathetic oneness with nature. For some traditional peoples, harming the land or killing any animal was like hurting or wounding one's own mother. Thus "primitive" and prehistoric man and woman saw themselves and all other animals as her children. By praying to the creative and benign powers of nature, humankind at one time lived in peace with themselves and with the earth itself.

The contrast between Rousseau and Gandhi's view that the natural life is superior to the sophisticated existence offered by civilization is important for any moral philosophy. As is well known to any student of history or culture, there have long been two opposed ideals of the good life. In ancient Greece, the Athenians with their democracy, drama, sculpture, architecture and groundbreaking philosophy already referred to rural Greeks as barbarians. In Renaissance Europe, the ideal was that life should be rich with religious art and architecture, great literature and creativity so that one's personality and intelligence find expression through multiple attainments.

Note the essential significance of this ideal. It implies that life is for exploration, experimentation, invention and creativity. The aim was to attain the realization of one's aptitudes, talents and intelligence. All these activities characterize the morphogenetic ideal of life.

By contrast, during the same Renaissance, there was the contrary ideal of leading a life of religious devotion, simplicity, and deep, spiritual morality. The simple life was to be realized through self-denial, self-discipline, and obedience to one's Christian conscience. The individual must shun all sins and vices but pursue the virtues of both Greek antiquity and Christianity. In sum, one should dedicate life to fulfilling a spiritual destiny. In essence, it was the symbiotic ideal of life.

By juxtaposing and contrasting the creative and moral ideals of these past civilizations, we discover how cultural and religious ideals express the essential, morphogenetic and symbiotic nature of humanity. Furthermore, we begin to glimpse the moral significance of the concept-metaphor, the creative conscience.

One of the renown interpreters of evolution was the English philosopher Herbert Spencer (1820–1903). He interpreted evolution as proceeding in the direction of an equilibrium, "a balanced combination of internal actions in the face of external forces tending to overthrow it." (Lillie, p. 188)[1]

The idea of balance between opposing tendencies is as ancient as the Old Testament and Plato's moral philosophy. Seen as a virtue, justice advocated the harmonious integration of various viewpoints. In Aristotle, the good was the mean between opposing vices. Therefore, the tendencies in human nature should be harmonized to better ensure morally right decisions and actions.

Applying the concept of equilibrium to evolutionary ethics has interesting implications. It does not simply mean maintaining the status quo. Rather, it implies the creation of new, morally worthwhile actions.

The theory proposed by *The Creative Conscience* has emphasized that morphogenetic creativity and symbiotic integration interact dialectically. Periodically, the activated first phase is complemented and mentored by the second phase. This interaction goes on cyclically throughout the life of a creature. Throughout evolution it is probable that periodic activation and quiescence take place.

Indeed, the earth itself tilts to provide seasonal variation, and the 100,000 year orbital changes (becoming more circular, then more elliptical) initiated and ended ice ages. In addition, it wobbles on its axis every 26,000 years, changing when winter and summer occur. Obviously, over eons of time, such periodic events must also have influenced the cycles of life on earth.[2]

As regards any moral philosophy to be extracted from the biogenetic facts of evolution, consider the following. In forming moral judgements, it is probable that morphogenetic creativity also exerts its influence on the symbiotic conscience. In other words, morphogenetic openmindedness modifies the judgement of conscience so that it alone is not the final arbitrator in ethical matters.

Moreover, morphogenetic discoveries, new knowledge, and new perceptions in human morality need continually to be acknowledged. Therefore, our inborn creativity acts to transform judgement to a more sophisticated, sound, and holistically complete symbiotic conscience.

ΣϽCg

What moral philosophy are we to extract from Darwin's theory of evolution based on natural selection and the survival of the fittest? In Darwin's theory it would seem that survival depends on the successful propagation of one's own species and the destruction of rival ones. (Lillie, p. 190) That hardly seems a safe ethic for humanity to follow.

To be sure, other theories of evolution have emerged in the past century and a half since Darwin published *The Origin of Species*. (1859) Lloyd Morgan's "Emergent Evolution" is mentioned by Lillie. To Morgan, various stages of evolution result in the emergence of something new—unforeseeable and unpredictable. "The emergence of life from nonliving matter and the emergence of mind from living matter are two of the most striking examples of the appearance of what is new and unpredictable in the course of evolution." (Lillie, p. 191). Morgan believes such emergences are determined less by mechanical causes than by ideals. (p. 191)

This insight has broad implications. For instance, consider the continual emanation of ideas we have all of our lives. They are not simply to be explained by "causes." How is it that intellectual, scientific, psychological, and creative "breakthroughs" characterize the history of civilization and culture? Causal explanation is clearly illogical in such instances.

Then again there is the case of ideals. They galvanize and motivate men and women's lives. Obviously, ideals visualize some future state of perfection or completion. Hence ideals themselves may be guided by some deeply embedded teleological instinct in humankind, which may well have had its influence on evolution.

All animate nature seems to sense it has some purpose in life intrinsic to itself. So it may be possible that in humankind the dialectic between morphogenesis and symbiosis has in itself a teleological purpose, namely to foster intelligent life capable of survival and of improving the species.

In the light of science, knowledge and the humanities, it becomes increasingly difficult to "have faith" in mechanistic, causal explanations of life in any of its forms. One fact we do know with relative certainty is that life is real and that the evolution of intelligence probably is the ultimate purpose of animate nature. In contrast to Blaise Pascal's famous bet on the omnipresence of God in the universe, we have at least biological evidence that intelligence in myriad degrees and forms is omnipresent on earth.

Henri Bergson undertook to explain evolution as the consequence of a "vital urge" or creative impulse. In other words, for him evolution is neither mechanical nor teleological (in the sense of Providence guiding humanity), but creative.

It is Lillie's conjecture that if creativeness is indeed characteristic of evolution, then the process may lead man to some kind of creative morality. He mentions Prof. L.A. Reid, who finds the motive of love to be the moving force in such an evolution. To be sure , Reid has in mind the ancient Greek concept of agape, which as a moral principle extended love to all humanity.

On the other hand, the Russian theologian M. Berdyaev believed that creativity itself is a kind of ethic. In creating something never before seen in the world, the creative act implies freedom. Moreover, each individual is responsible to create himself or herself since such moral activity emanates from the depths of one's own conscience. (Lillie, p. 194).

Thus my own account can be expressed in another way. Not only is humankind morphogenetically creative and symbiotically conscious of humanity's need for compassion and aid. In the depths of our evolved human nature, we discover manifest nature's creative conscience. Not only is that conscience in nature adroit in evolving the most effective plans and exquisite designs in all forms of life. It also aims at evolving in humanity a moral nature that will be equally resourceful and skillful, yet just and humane. In some future millennium, morphogenetic perfection and symbiotic integrity may truly have evolved a universal moral conscience.

It is clear then that without the expression of one's creativity and the realization of one's own self-worth, we cannot evolve a truly compassionate and creative conscience of our own. As the true artist endeavors to create a perfect work, the individual needs to perfect personality or soul. A form of morality in its own right, human perfection must involve creative and conscientious self-surpassing.

The concept of self-realization is implicit in the description of the moral life. As observed in nature's capacity to perfect the design of plants and animals, for humankind self-realization affirms the need for the individual to mature and cultivate the self.

However, beyond being a laudable injunction, how is the human being to undertake his or her moral evolution? The answer is to exercise one's inborn nature. In other words, we must seek a reasonable balance between creativity and conscience. Since our nature is a dialectic of polarities, two caveats are in order: (1) our instinct for freedom should not be allowed to degenerate into anarchy or insanity; (2) conversely, we should not allow our conscience to be enslaved by inflexible orthodoxy or by unquestioning faith in the judgement of so-called "superiors." Put another way, every self-respecting individual needs to mentor his own creativity and freedom by his or her conscience, and that self-same conscience needs to be monitored by creative open-mindedness and open-heartedness. Rather than devolve into a judgemental and moralistic conscience, a true morality should correct excesses and anarchy with tact, sophistication, and humane justice. To be moral means to act both justly and charitably. We need to use both creativity and conscience to make our life decisions.

As is well known to students of ethics, the ancient Greek philosopher Aristotle defined the aim of the moral life to be *eudaemonia*, by which he meant the realization of one's physical and mental capacities to achieve excellence. Indeed, such commitment was regarded as "virtue."

The German philosopher G.W.F. Hegel (1770–1831) argued that the universe itself was undergoing a spiritual evolution actuated by a dialectical process. (Hegel believed the realization of a concept was such that thesis + antithesis yielded a new synthesis.) In Hegel's view, throughout human history, man's capacity to think evinced dialectical growth and development. Thus the manifestation of physical and psychological change in an individual's life as well as the evolution of species strengthen the argument that self-realization is a moral injunction to be obeyed by every mature person.

The ethical idealism of T.H. Green *(Prolegomena to Ethics)* believes human nature is to be characterized by a "spiritual principle." To be sure, this is a very comforting point of view kin to the religious belief in spiritual powers roaming the earth since the beginning of human time. Of course, to date we have found no sure means to substantiate this hope.

On the other hand, we have abundant scientific evidence as to our biological heritage. As children of nature, her bio-processes have imbued us with intelligence, creativity, and conscience. Mystically speaking, there may be a "spirit" in Nature akin to the concept of the "Supreme Being" or perhaps the "Supreme Becoming." Indeed, if nature has Intelligence, that Intelligence may incarnate Spirit. (cf. Bergon's "vital energy"). When one witnesses the germination of the simplest seed or the birth of a bird from an egg, it is impossible not to wonder about the Power that brought them into Being and future Becoming. Clearly, they are the elemental expression of some ineffable Form in Nature.

T.H. Green also thinks man's "power of looking forward to the realization of an idea" due to an inherent spiritual principle. While this sense of our "spirituality" may be very comforting to many, Green's version is not necessarily a realistic interpretation of humankind's power of foresight. A more mundane but honest explanation would be that the individual's ontogenetic development throughout a lifetime is what enables us to foresee the direction of our lives. It also explains how stage by stage (childhood, youth, adulthood), we gradually acquire a more mature glimpse into the meaning of life itself. On the other hand, the ability to foresee mankind's needs and forestall human misery may be more than practical foresight. Perhaps this spiritual aptitude comes from the moral evolution of our conscience.

For better or worse, we evolve by integrating our lives through pursuing a higher or more worthwhile purpose; or we devolve by allowing ourselves to become victims, chasing senseless desires that leave us with a sense of personal worthlessness. The choice is ours: to evolve by integrating the energies nature has so freely given us, or to devolve by letting our true identity disintegrate.

Although mankind has yet to understand the source and meaning of life itself, we are allowed to have a measure of confidence that humanity is endowed with sufficient, inborn intelligence and sense of purpose to emulate what we imagine to be the Supreme Power in the Universe.

It is important to understand the need for a new morality that can endure and transcend future centuries. By exercising our aptitudes, talents and forms of intelligence, we become more inner-directed. As a consequence, we become secure enough to help others in need, generous enough to help them help themselves.

Moreover, by conscientiously endeavoring to harmonize all our own capacities, we learn to respect ourselves for our inner strengths, for our ability to develop further our Nature-given gifts, and to achieve a level of excellence in all that we do. In other words, a truly moral man or woman learns to consolidate and integrate as many natural endowments as practicable. Put another way, individuals need to explore their potentials for creativity and to cultivate their conscience. For only the whole person can truly help others realize the worth of their own lives.

In the context of our more advanced knowledge of nature, we can attain happiness in two ways: (1) in the lifelong use of imagination to generate ideas and creative projects, and (2) in the stage-by-stage development of one's symbiotic conscience. To effect such a mini-evolution, one must undertake (a) to perfect the design and complete the purpose of whatever we plan, and (b) to educate our moral sense by accepting responsibility for the betterment of humanity near and far. Thus personal happiness becomes a reality when we foster our inborn gifts and nurture our emotional intelligence. Humanity needs more than the expression of our sympathy. We need to help humankind by whatever means we can create.

Of course, each day brings with it responsibilities in school, work, or family. Whatever one's job or profession, there often seems little time for creativity or for thinking about the needs of others. Anyway, don't we really believe "Charity begins at home?" The average individual seems at times unable to summon enough energy to take on a program of self-realization or to work out a moral philosophy involving humanity.

While that is generally true, we all know how much time we waste on trivia and meaningless activities. Although all need some form of recreation, just how much do we do that actually recreates us or refreshes body, mind and heart? How often do we actually listen to the needs of body and heart? Do pain killers actually meet the needs of the mind? Do we once in a while wake up to the fact we have become a stranger to ourselves? What happened to our idealism, our love of learning, the good feeling that came from the conversation between friends who trust one another? Does the work we do satisfy our sense of challenge and creativity, our need for a sense of meaningful accomplishment! Or are we sacrificing everything for the money?

If we can't hear any answers, then what are we doing with the one life we have? If there is nothing but silence to those questions, then maybe it's time to figure out who you are and what you want to do with the rest of your life.

On the other hand, you may be one of those human beings with more ideas than you know what to do with, or you never go far with the ideas you have had. Maybe you have started many projects but rarely brought them to completion. Or perhaps they remained stillborn because you didn't have time to properly develop them, or your conscience was too critical and you aborted the idea by attacking it prematurely for its weaknesses and shortcomings. The idea was then discarded as not being worth serious commitment or sustained development.

To offset such tendencies, it is necessary to give nature's morphogenetic creativity time to evolve your seed ideas before judging them. Sometimes, it is important to follow willy-nilly an idea which can lead you, with a bit of luck, to a new design or a revealing pattern. Sometimes, a tiny idea can open up vast vistas of perception or dimensions of universal meaning hitherto unsuspected.

Then after a "reasonable interim," it may be time for your symbiotic conscience to step in to assess, at least in a preliminary way, the sense and application of your core concept to the real world of imaginative works. Will the idea, theory, or invention "fly"? Of course, if you are working in a field in which you have already a certain experience, that would mean the symbiotic conscience has accumulated relative expertise in evaluating your new creation.

So, in sum, once again, morphogenetic creativity and symbiotic conscience interact to synthesize a more perfect, more complete creation much as, over millions of years, Nature herself has evolved the successful species that populate our living world.

The Greek word for virtue meant any form of excellence. To the ancient Greeks, the four cardinal virtues were: justice, courage, self-control, and wisdom. For modern man and woman, I would add "Just as you educate yourself to develop practical skills or acquire a profession, you need also to educate your conscience."

These qualities of moral life raise an all important question as to the ethic proposed by *The Creative Conscience*. Thus far we have argued that Nature itself is animated by morphogenesis and symbiosis. In so far as these processes are inborn in human nature, they should be consciously nurtured and fostered.

For instance, in so far as morphogenesis motivates our innovations and inventions, we should pursue these inclinations and cultivate them. Thus creativity should be applied to our everyday and professional world, but also to educate our talents, emotions and intelligence.

On the other hand, symbiosis in nature is responsible for perfection and completion of design. It perfects and completes the roots, stems, leaves and flowers of plants. It integrates cells into tissue, organs, and life support systems. It forms life into species.

It has a corresponding responsibility in our sentient-mental life. In human nature, symbiotic activity seems to pursue the purpose of maintaining a balance between strong emotions and self-restraint so as to ensure soundness of body and mind. Yet our symbiotic nature is not limited to a sensible conscience obeying the injunction "nothing in excess." In addition, our symbiotic nature aspires to perfect the mind through knowledge and cognition and to complete the education of the psyche by guiding it to incorporate as much experience of life as possible.

As far as mental dexterity is concerned, symbiosis implies learning skills of coordination and subordination to prioritize goals and objectives. Moreover, symbiosis urges us to combine our heterogenous interests into integral units of understanding and action. Symbiosis coalesces the disparate into symbolic meanings. It concentrates our energies to focus one's being on becoming. It guides the transformation of the psyche into higher stages of consciousness.

Furthermore, our symbiotic nature empowers us to reach out sympathetically to our own kind, to practice compassion where it is due, yet it also demands we carry out justice for all, including ourselves.

In scrupulous coordination with morphogenesis, symbiosis ultimately evolves the creative conscience, which actualizes all the above attainments. To be sure, the creative conscience monitors and moderates one's conduct in a world dedicated to the ideal of equality and equity for all humankind.

On the other hand, our future evolution as a species should be such that the creative capacities of humankind be mentored by our conscience and our conscience be moderated by humane comprehension. A sensible understanding of our antipodal nature should teach us to reach balanced judgements. Creative decisions and actions during our lifetime should countervail, counterbalance, and compensate for the distress, catastrophes, and tragedies of the millennium we live in.

There is yet another mental characteristic of our biogenetic nature. Initiating analyses is a primary function of morphogenesis. Via exploration of the known and the unknown, by discovering similarities and by distinguishing differences, morphogenesis analyzes the facts, phenomena and events of its environment. It concerns itself mainly with the finite and the definite. (Note the similarity to the intent of Plato's Dialogues, in which Socrates distinguishes the lesser truths from the greater. Also note the parallel use of negative definition in modern expository writing where a topic is first defined for what it is not and then defined by what it actually is.)

By contrast, symbiosis aims at synthesis. It undertakes to encompass and integrate as much knowledge as possible—data, detail, uniform phenomena, and patterns of meaning. Hence symbiosis is a kind of positive definition much as Hegel's dialectic, which aims at a comprehensive explanation. In other words, based on investigation and accumulation of data, Hegel undertakes to discover generalizations and incorporate all-embracing inferences to his history of humanity. His conclusions integrate his knowledge into universals of meaning. Thus his syntheses corroborate his analyses of historical events and vice versa his analyses substantiate his syntheses.

Let it be borne in mind that each polarity (morphogenesis and symbiosis) has its turn in actuating and actualizing our psychic, rational and moral activities, yet ultimately they work synergistically. Periodically they establish or re-establish homeostasis, but cyclically they work in concert to perfect our moral nature and to further the evolution of our sentient intelligence.

Nature's polarities act via opposing tendencies: divergence and convergence which have influenced the course and development of human cultural history. Similarly morphogenesis and symbiosis account for patterns of behavior and the evolution of ethical systems.

The beginning of the Third Millennium finds itself largely secular in attitude with vestiges of faith and orthodoxy failing to sustain us or our societies. In a period of history that daily appears more threatening, life often seems senseless and without any viable morality. To be sure, intellectuals and professionals still have a secular faith in man's know-how, technology and inventive genius. Well and good, as long as that ingenuity does not invent ever new "smart weapons" for annihilating whole populations, including our own. The satellite communications systems in place should notably encourage unity and understanding among humankind. However, the military misapplication of human creativity in the worship of death may well lead to the extinction of the human race.

The frail hope of *The Creative Conscience* is to awaken the emotionally intelligent of this world to a moral philosophy which is sane and sound in having its roots in the Science of Life. If any Power in existence should be worshiped, it should be life itself. If there is any possibility of humankind pursuing a nobler destiny, it is through understanding the higher Power in Nature. It urges us to cherish life. It appeals to us to use our creativity to promote understanding and ensure justice and compassion for all. It advises us to be humanely human.

The great religion of the earth should be the love of life and of all who share this planet with us. To glorify war is to surrender to the worship of death. Let there be peace for every child of nature. Let us help all human beings realize a life worth living. Let us teach one another we are not the enemy.

We are the friend, the cousin, the brother, the sister, the father, the mother to all who need us. We are of one blood. Let us educate each other to respect our kinship. That may not happen in this millennium. But who knows what the next 2,500 years will bring?

11

THE CREATIVE CONSCIENCE AS HUMAN DESTINY

Introduction:
Conflicting Interpretations of Evolution in History

Darwin repeatedly used the term mechanism to explain how biological nature acts and functions. This mechanistic analogy clashes directly with the detailed evidence he amassed to prove his theory of evolution. Moreover, physicists and chemists have argued that nature provides ample evidence for the physical, chemical and electrical conversion of "energy" as the ground of being for evolution. Yet any attempt to interpret the metamorphoses of life forms, on the basis of matter and energy alone, is blind to the actual biofacts. Any argument that holds animate processes to be mechanically determined and capable of explanation by mechanisms has used a misleading analogy contrary to the facts.

Rather, mutation and variation among life forms indicate that cellular, organic, and somatic experimentation display an intrinsic capacity to seek out more effective means of survival. In other words, the metamorphosis inherent in life forms, from the "simplest" to the most intricate and complex, clearly reveals a purposiveness in animate nature. Whether this process is entelechal or teleological or "simply" the evolution of life through distinct stages of completion and perfection, all species that we presently know manifest a similar history and direction.

When Darwin showed deference to Newton's mechanical interpretation of the universe or to Galileo's mechanism, he revealed his yearning to make biology an absolute science. He hankered after the certainty of physics and chemistry. However, the salient fact in any mechanistic or mathematical description of the universe is the fact that forms of life in vital nature transcend all chemical formulas or thermodynamic principles, as valid as such explanations may otherwise be.

Life may have emerged originally from conditions described ultimately by physics, thermodynamics and chemistry in terms of electricity and plasma. Yet scientists are still unable to explain what electricity is or why it exists. Nor can they say anything more about plasma than it is thought to be a collection of charged particles with positive ions and electrons interacting most often in distant stars. How do such explanations help us understand life on earth?

On the other hand, a biologist may be eager to regard the dynamic nature of the stars as replicating the metamorphosis and integration of their elements in some sort of endless experiment with existence.

Will we ever know the sense and substance of the cosmos? At least twentieth century biology has given us a more earthy, useful science focused on understanding life and humanity. This advanced biology goes beyond all purely mechanistic, causal explanations of existence. Microbiology and ecology have ascertained facts and phenomena which have released us from the blind causality and coercive mechanism that seemed to rule over life. The physical sciences made the human being seem a victim of relentless, inhuman forces.

The great mistake of nineteenth century biological mechanists was to extend the mechanical analogy too far as René Descartes himself had done in the seventeenth century. Life forms are not bio-machines or androids. The human being represents a more sophisticated order of nature.

Moreover, life needs no analogy to explain it. Life explains itself via its natural phenomena. That is, it defines itself by how the individual life form deals with the limits and challenges of the environment. Ultimately, as a viable organism capable of survival, the form must deal with the limitations within itself.

This juncture between environment and self decides its death or survival. To meet the threat, the individual and species must undergo change to improve their capabilities and must consolidate their evolved strengths. Thus life forms become more efficient and effective in survival.

The argument presented thus far in this book is simply a refinement on Darwin's theory of evolution. Rather than nature being, in effect, primarily a struggle for survival, life endures because of the individual capacity to evolve its own intrinsic morphogenesis and symbiosis, which enables the species to overcome the limitations posed from without and to master their weaknesses and limitations within.

Due to the contributions of ecology and microbiology, a more complete theory of evolution is being realized. Furthermore, anthropologists, psychologists, and philosophers are becoming aware that a new definition of human nature is needed. The biological evidence reveals two universal processes in nature, morphogenesis and symbiosis. In humankind, these have evolved into creativity and a nature-born conscience.

The New Reality

Twentieth century ecology introduced us to a new reality. It was the discovery in nature of a great number of mutually beneficial relationships between different species that enabled them to survive. These discoveries delighted us because, hitherto, we had been largely blind to their existence. On the other hand, microbiology brought into focus the reality of the microworld. Inhabited by unimaginably tiny forms of life, it awed us by allowing us to see into the cosmos of the infinitely small.

In the past century, two universal processes were discovered. From the cellular dimension of bioactivity through all stages of growth and organization, these processes account for our individual ontogenesis as well as our long evolution as a species.

The first process was called morphogenesis, which we have already defined as the formation and differentiation of tissues and organs initiated by microscopic cell activity, but continued until all stages of biological transformation are completed.

The second process was named symbiosis, which manifests itself in three spheres of life. (1) Throughout nature there are many examples of mutualism between distinct species. (2) The human body itself, as well as the body of all living creatures, reveals it to be an integrated organism. (3) In all forms of life, every living cell has its own integrity and each, in turn, unites with others to organize into tissues, organs, and life-ensuring systems.

Before we discuss how these twentieth century discoveries offer us a new reality, let us describe the old reality based on Darwin's theory of evolution. As explained earlier, Darwin's thorough and comprehensive research into the habits and habitats of species remains to this day worthy of our respect. Over the past 150 years, Darwin's theory has inspired worthy pursuits into various fields of study.

History has shown that his argument that we descended from simian and hominid ancestors affronted theologians and churchgoers. However, his scientifically-sound realism taught us hard, if difficult to accept, truths about human nature. The fact that we originally emerged from the sea, became land animals and even monkey-like men and women surely gave "civilized" man a harsh dose of reality. We learned that our struggle for survival likely shaped us as an aggressive species determined to stay alive, regardless of any other creatures. Or so it seemed.

The theory had a negative impact on society. It seemed to validate ruthlessness in human relations. It turned some of us blind to the need for compassion. In fact, the theory seemed to accept violence and homicide as our heritage from nature. This interpretation not only condoned vicious laissez faire in business and commerce but also encouraged the abuse of human rights and the exploitation of human beings. The powerful felt no fear of being held to account for their selfishness or indifference to the fate of whole populations, whose starvation wages and appalling living conditions forced the poor to be treated as subhumans.

Indeed, "fang and claw" seemed the nineteenth century symbols for Darwin's *Origin of the Species* and *Descent of Man*. Little did he suspect that his patient passion for discovering nature's secrets would later reveal how ineffective our "civilized morality" had become. The greed, the manipulation of the poorly educated masses, the insatiable hunger for power seemed to confirm the fact that mankind actually lived by predatory instincts. This dismal view of our biological heritage would surely have disheartened the modest, conscientious soul of Darwin. Yet the vulgarized version of Darwin's theory became prevalent.

Curiously, for millennia before Darwin, existence was regarded as the scene of a war between powers of good and evil. The diabolic and demonic seemed ever ready to destroy the unwary, helpless and defenseless. Earthly life was uncertain because malign powers were lurking everywhere to devour the innocent and to punish the passionate. Like nature's predators, Satan's beasts were waiting for all of us. In other words, ignorance and superstition superimposed age-old fears and nightmares on the theory of evolution. Thus, for a while Darwin himself was demonized.

Then there were those pseudo-scientific interpreters who superimposed their own version on the theory to justify their indifference to morality. After all, success in life belonged to the "survival of the fittest." This meant that the shrewdest, the most deceitful, the most cunning, the cruelest would be the survivors. In fiction and newspaper articles, in crime/spy stories, and in television documentaries, it was the toughs, the heartless, the killers, the villains who seemed elevated to hero status. They were the real protagonists and the masters of fate in an amoral, predatory world.

Let the old reality pass. Let the past of cunning, craftiness, and duplicity in human relations pass away. Let the self-serving, egomaniac, and demagogic distortion of Darwin's theory succumb to a sound account and trustworthy exegesis of nature's inner laws and true lessons for humankind.

※

It is time to consider the human implications of morphogenesis and symbiosis.

Unexpectedly, twentieth century ecology and microbiology predicated a new life philosophy for the Third Millennium. By investigating nature from two distinct perspectives—the infinitely great and the infinitely small—biologists discovered that existence has dimensions of meaning which together possess a unitary significance. Ecology found that the earth's biosphere, and microbiology discovered that the tiniest cells of life, each manifested virtually identical bioprocesses. Nature was one.

Each study in its own way perceived that morphogenesis and symbiosis were manifest in each realm of reality, the macro and micro spheres of the earth. These processes were responsible not only for creating the tiniest and largest life forms but also for consolidating the relationships between all living things, large and small, in the biosphere of land, sea and air.

In addition, these processes influenced individual transformations over a single lifetime as well as affected the evolution of species across the ages.

Moreover, each process acts in characteristic ways. Morphogenesis is made evident in the way creatures explore, discover, experiment, and create. These visible activities obviously originate in and evolve out of a life form's innermost being. As such, morphogenetic activity displays the creature's adaptability and resourcefulness when they are needed to survive.

On the other hand, twentieth century microbiology has shown us how symbiosis acts in other characteristic ways. A human life form undergoes integration from cell to tissue to strands of muscle; from growth of bone to the structuring of the skeleton; from the organization of organs to the completing of life support systems.

Physiological and organic changes must also influence cerebral evolution. In human beings, our inherent morphogenetic intelligence manifests itself by the ways it deals with all manner of contingencies; its teaches us to defend ourselves, to preserve health and well-being. Moreover, the history of humankind provides ample evidence of the many things we have invented to express our creativity.

Similarly, our cerebral evolution is manifest by our inherent symbiotic intelligence. Here the human mind evinces its aptitude to coordinate, combine, and consolidate its experiences by resolving practical problems, perfecting designs and plans, and by coming to terms with existential truths and meaning. This form of human intelligence senses that survival of oneself and one's descendants depends on two correlated efforts: (1) to connect, condense, and unify into meaning what we learn from life; and (2) to define our destiny by pursuing a lifetime purpose of our own.

It would seem that nature itself is urging humanity to understand the ultimate significance of evolution. Through ecology and microbiology, nature has made it clear that morphogenetic and symbiotic intelligence are the reason for evolution and for the actual survival of life on earth.

Moreover, it seems self-evident that human society should emulate Nature's Intelligence. Ideally, societies of skilled, educated, and ethical individuals working in tandem should have one mutual purpose: to aid one another. If many societies join together in this effort, a new kind of world civilization could evolve. Eventually, the individual conscience would be dedicated not only to one's own ambitions, to family and country but also to the needs of all humanity.

The Question of Extinction

The term *extinction* defines the act of making a species extinct, the fact of being extinct or the process of becoming extinct. An extinct species no longer exists. It has been eliminated because it was in some way inferior to others of its own kind or it was displaced by another species. Darwin explained this phenomenon in nature as being due to "the struggle for survival" and nature's selection of the fittest.

However, a contemporary understanding of extinction may go beyond the somewhat mechanistic interpretation of biological processes. Morphogenesis may well provide a more accurate explanation since the micro-process accounts for useful, life-enhancing innovations that produce variations and mutations. In as much as we have affirmed there is such a capacity in nature as morphogenetic intelligence, that may be the key to survival or extinction.

What would happen to any species unable to actuate inner adaptations when such permutations would increase the chances of survival? Obviously they face extinction.

Yet the earth itself has changed many times since land masses, sea and breathable air became the more or less stable condition of the biosphere. In other words, all such changes would necessitate some form of metamorphosis in all living creatures. To survive, it became necessary to adapt, to alter, to transmute their cells or genes and to actualize certain physiological transformations. Thus it is probable that this capacity for genetic creativity must have been the essential factor in the survival of the individual and of many species.

The failure to change when change is needed shows a state of stagnation. By being unprepared to change when environmental conditions cry out for it is another aspect of extinction.

Of course, some species continued to survive because the immediate environment hardly changed over millions of years. (e.g., the shark or crocodile). Ever since their early evolutionary successes, the further evolution of some species became unnecessary.

Otherwise, land animals exposed to natural disasters, drastic changes of climate, or the encroachment of other species had to originate defenses and mutate. The more severe the changes, the greater the urgency to accelerate their evolution; the more mature their instinctive self-knowledge, the more capable they became in generating the changes required.

Hence we have here an alternate explanation for species extinction. The failure to initiate cellular, genetic, physiological or ontogenetic innovation could account for the extinction of individuals and species.

Small animals provide us with an example of survival against all odds. It is probable that, because of their size, they had to develop their instincts in ways distinct from larger animals as wolves, bears, moose, reindeer and the like. It seems that many small animals as rabbits, beavers, squirrels, chipmunks and the like acquired foresight which enabled them to survive. Hoarding skills got them through seasons of want, such as harsh winters. To be sure, their foresight was based on hindsight learned from what happened in the past. So it should be noted that foresight and hindsight, like creativity and conscience, taught both man and animal sensible ways to survive.

Yet humankind has learned to use these instincts with some caution. Anthropology has given us examples of societies and cultures which largely abide by the achievements of the past. They mainly obey the tried and true, yet at times they find that the most highly developed hindsight alone is not enough to survive. Sometimes the lessons of the past no longer apply. (Many of today's traditional peoples are experiencing great anxiety at the extreme machines, weapons, and "miraculous" communication systems of the technologically advanced nations.)

Indeed, in many tradition-bound societies, the inability to use foresight seems to threaten their survival as individuals and as cultures. This is one source of the hatred for the wizardry of the West. Yet if such societies fail to learn the value of foresight to prepare for the inevitable future, the individual and his culture will remain lost in a world he no longer recognizes.

However, eventual human extinction may be the consequence, not of a "struggle for survival," but of the confrontation between conscience and creativity. To be sure, conscience is represented by mores, customs and traditional ways of doing things. Creativity is represented by the need for fresh experiences in life, the opportunity to use one's imagination, the capacity to invent something new, and the desire for freedom.

Yet when conscience is too strict, it suppresses self-expression; it ridicules the new; its discourages appreciation of cultures outside one's own; it forbids thinking of life as freedom; it disdains what it does not understand. Gradually something natural in the individual dies. His morphogenetic intelligence has been mutilated, mangled, nearly extinguished.

On the other hand, if creative freedom degenerates into antisocial behavior, self-abuse through drugs or alcohol, the disregard for the rights of others, the disrespect for elders, the practice of excesses of all kinds, then so-called creative freedom has become anarchy, immorality, and a kind of suicide. Conscience has been scorched, smothered, suffocated. The individual's intelligence has been extinguished. He has degenerated into something inhuman.

In sum, individuals, societies, religions and races may go extinct because their creativity has been enslaved by conscience; or, conversely, conscience may deteriorate to the state of savagery before the rise and evolution of humane humanity.

These thoughts lead us to both psychological and social considerations. Psychologically speaking, there is a need for every individual to harmonize his capacity to create and the requirements of conscience. As such, introspection or meditation may inspire the individual to become more self-reliant and resourceful. Each of us needs to be guided by our symbiotic intelligence, which urges us to do two things: (1) to consolidate and integrate our personality through honest self-knowledge, and (2) to complete and perfect whatever we undertake in life. Following such guidance, we will come to know the purpose and meaning of our lives.

Socially speaking, we may be committed to upgrade our surroundings or to improve relations among all members of society. These are opportunities that both challenge our ingenuity and test the sincerity of our conscience. A balance between creativity and conscience would ensure that our life-worthy activities would help ameliorate the human condition. In any event, let our aptitudes and talents aid our fellow man and woman to find new faith in life.

As regards the question of extinction, there is reason to believe that nature's innermost modus operandi is the most efficient and effective means of guaranteeing survival. Through morphogenesis and symbiosis, all earth's creatures evolve and perfect their chances of leaving progeny for the future.

By the fact that human nature itself is a progeny of Nature, we come to recognize that we, as individuals and as a species, will be responsible for our own survival or extinction. By sufficient foresight and hindsight, creativity and conscience, those of us who survive will deserve to live on.

The Argument Thus Far

1. This book has presented a new interpretation of human evolution, intelligence, conscience and psyche.

2. This new interpretation is based on nature's polarities: morphogenetic creativity and symbiotic conscience both inherent in human nature.

3. Nature manifests a Creative Conscience, and human nature focuses the intimation that a lifetime should have an ultimate meaning in terms of Nature.

The Dynamic Polarity of Human Nature: Human Intelligence

What is intelligence? It is not a thing or an abstract concept. It is more than a potentiality or capacity. Intelligence is an activity, or better said, a process. It transforms the unknown into the known; it integrates what is new with the already learned. It passes from ignorance to knowledge. The process is effected by morphogenesis and symbiosis.

What is learning? It too involves morphogenesis and symbiosis, which together bring about the transformation from ignorance to knowledge, the transmutation of mind that we recognize as learning.

Morphogenesis has a practical role in seeking knowledge and a biogenetic purpose in assuring survival. Its responsibility is to learn all it can about the environment and the mind itself. Ultimately, it is attempting to define its own function.

Symbiosis reinforces, defines and refines the role of morphogenesis. It judges new incoming perceptions as to their survival value and practical utility in the light of memory and experience.

Morphogenetic intelligence interacts with symbiosis so as to understand more fully what it perceived through the senses and inferred from the mind's intuitions. In effect, morphogenesis learns from the past experiences stored in our symbiotic intelligence.

The function of morphogenesis is to satisfy its curiosity, to explore and to experiment whereas the function of symbiosis is to coordinate, to coalesce and consolidate the findings and experiences of its partner. Thus they knit together the newly learned with the already known.

In this way, human intelligence broadens and deepens its comprehension of self and life. This process of perception/apperception enables the mind to reach stages of viable knowledge and enables the conscience to evolve new forms of wisdom. The overall process proves that man's mind epitomizes the Creative Conscience of Nature itself.

In contrast to our public education, and distinct from learning a trade, vocation or profession, how did natural intelligence educate us?

Long before apprenticeship, learning practical skills and formal education, the human mind had learned how to learn. It did this by instinctively satisfying its own curiosity, by being creative and by searching to understand life.

By exploring the life within ourselves and by experimenting with the raw materials of the world around us, human intelligence taught itself to comprehend what survival meant. In other words, the mind's capacity to integrate experience gave evidence that human intelligence was capable of creating knowledge through the consolidation of life's lessons.

Obviously morphogenesis and symbiosis, incarnate in flesh and bone, account for our psyche-mind. These complementary functions stimulate self-actualization and self-fulfillment and/or self-realization and self-perfection. These body and mind processes are responsible for the development of the individual's intelligence and the evolution of hominid into Homo sapiens.

The Morphogenetic/Symbiotic Conscience

The fact that evolution has endowed humankind with a natural conscience, or a conscience born-of-nature, allows us to describe human conscience in biological terms.

As we have already learned, morphogenesis and symbiosis are universally manifest throughout nature. Both processes are present in cellular, insect, plant and animal life.

At the bacterial and viral level of life, morphogenesis and symbiosis are active, the former in the capacity of self-transformation and the latter by adaptation to their host, which processes together guarantee survival.

Insect life can metamorphize itself in color, size, shape, and mobility, and by their clever camouflage are able to adapt to their surroundings. This accomplishment takes morphogenetic ingenuity and symbiotic accommodation. Again, the change serves the purpose of survival.

Plant life establishes a symbiotic relationship with sun, air, rain and soil and is morphogenetically capable of producing its own food. By their seed, fruit and nut, they entice birds and animals to ingest them and thus disseminate their descendants far and wide. Also, plants invent other ways of dispersing their seed by water, wind and insects. Apparently plants have had a timeless symbiotic relationship with their environment, its conditions and creatures. Moreover, plants seem quite capable of defense by manufacturing liquids that are poisonous to insects and animals that would otherwise devour them. They can also exude toxins to defend themselves against the encroachment of other aggressive plants. Evidently this capacity manifests their morphogenetic intelligence or consciousness.

To be sure, animal life has adapted eco-niches worldwide with different conditions, environments, and climates. Animals adapt themselves to various terrains, altitudes, vegetation, predators and prey. They are often able to camouflage themselves according to the environment and the season of the year. Each adaptation, requires morphogenetic adjustment or variation as well as a symbiotic sense of what will succeed.

As a special species, hominids survived via learning how to morphogenetically change raw materials (as bone and stone) into tools; (tree trunks and bark) into vessels; (wood and stone) into shelter; (plant fiber and animal hides) into clothing; (plant leaves, juices and roots), into medicine; and the like. Clearly, these are illustrations of morphogenetic creativity, and by the products serving specific uses, that shows a symbiotic consciousness based on utility. Anything made, designed or used for a specific purpose shows symbiotic conscience. Where the means fulfill a need or an end, there you have symbiotic practicality.

Put another way, wherever nature provided an example of interconnections, organization, structure, or design in insects, plants, or animals, there we find humankind applying its symbiotic understanding of nature to practical ends. Thus practicality required morphogenetic ingenuity based on symbiotic knowledge.

Another example of morphogenetic perception with symbiotic purpose is the following. Our hominid forebears butchered animals not only to roast them for food but also to use the bones and hides. Moreover, it may have taken a very long time, but eventually the most observant among them must have noticed the similarity of inner structure between men and animals. Studying the dismembered parts, they must have been impressed by the body's remarkable interconnectedness. Yet not until five or six centuries ago did we really begin the medical study of anatomy and noted the singular interdependence of all its tissues, muscles, skeleton, organs and brain.

Finally, we come to the body-mind which also evinces both morphogenesis and symbiosis. By its ability to respond to needs and emergencies requiring us to solve life-threatening situations, the mind is able to change attitudes, tactics, strategies and actions so as to survive or to overcome. This is the morphogenetic sphere of activity.

On the other hand, by its ability to coordinate, orchestrate, coalesce, and consolidate all the body's senses, organs, and life-support systems, the mind is able to actuate several stances. It may resist malice or withstand antagonism and aggression; it may shield itself against danger. It can maintain equilibrium and readiness in the face of hostility; it can combat predators. All these responses are based on symbiotic knowledge of the creatures living in the eco-niche.

With the realization that the human mind responds to inner needs and external requirements in characteristically creative and conscientious ways, we come to appreciate the fact that its morphogenetic/symbiotic functions and purpose are the biological basis for describing the mind as a creative conscience.

The body-mind, which actually defines the scope of human intelligence as a biological being, survives when everything within works in harmony. As the "center" of the self, the brain has an instinct to bypass mortality through our children. In other words, the mind aspires to immortality through leaving descendants who will beget descendants.

However, for those who do not want the lifelong responsibility of nurturing, caring for and educating progeny, there are other ways to try to "live on." As we know from human history, there have always been individuals endowed with an overriding sense of purpose. They have sought to pass on their intelligence, thoughts, feelings, theories, life stories, commitments to a cause, and philosophies by leaving behind remarkable works, endowments and monuments of all sorts. Such individuals leave behind something of worth that endured the ravages of time. We are grateful to them because they greatly extended our understanding of the many capacities of the creative conscience of humanity. Their messages survived death through the originality and moral sense of their thoughts. They were worthy to be appreciated by their descendants.

Yet beyond the wish not to be forgotten, the creative artists, writers, composers, scientists. engineers, and philosophers have rendered us an enduring service. By their creativity and conscientiousness, they provided evidence that all humankind are so endowed.

The fact that our morphogenetic nature and symbiotic conscience are truly the source of our intelligence is a rather startling revelation. Those who learned to tap this power, inherent in ageless nature, drew on the greatest force available to the human being. They educed the source of life itself.

This insight should make us aware of its importance to virtually all we know and do. If scientists, humanists, philosophers, and educators grasp this essential truth, that realization must influence the human sciences, the humanities, epistemology, knowledge per se, and the education of generations to come. It would seem that the creative conscience is a power greater than the atom itself in so far as it is under our control to further educate and humanize us.

The Morphogenetic Phase of the Creative Conscience

Morphogenesis in human nature enables us to accept change and to actualize change. It makes possible self-transformation and mental metamorphosis. It activated our hominid evolution and was responsible for our conversion from ape to man.

If we accept the fact that our species survived because humankind was able to realize its inherent potentials, this insight should have significant implications for our future. To realize anything implies coming to understand the essence of an existential situation. In addition, to realize means to set a goal, to pursue a purpose, or, at least, to explore an opportunity.

We explore the unknown to find out our survival chances. This habit of exploration likely leads to experimentation with the raw materials of the environment. Archeology has provided abundant evidence of experimentation in man's prehistoric artifacts and tools, their shelters and weapons. Even their methods of preserving food and cooking, their skills in hunting and gathering, their later agriculture and settlements all tell us they used creativity in their everyday lives.

Hominids and prehistoric humankind responded to circumstances and situations in various, imaginative ways. Whatever the problem, they used improvisation to contrive devices and traps, to concoct plans to hunt down game, made use of mud clay to produce pottery, fashioned all manner of shelters, and erected defenses against encroaching tribes. Moreover, they created trinkets, body painting, dress and ornaments to beautify the female or add to the look of virility of the male. But most important of all, of course, they invented language, in fact, about 1500 of them.

The external evidence is, more or less, all we have found at archeological sites. We know little of hominid dreams, imagination, ideas, or religious beliefs. We know only they respected the dead and prepared their bodies with clothing, adornments and their favorite possessions to accompany them in "the other world." Of course, cave paintings showed their extraordinary artistic ability. Yet these findings tell us that there must have been poets, myth-makers or philosophers among them. Hence death must have made them pause to reflect on the life we possess and what it means to be alive. Inborn creativity seemed the only viable answer to death. Our flights of imagination liberated us from our earthbound mortality.

We do know from world history that humankind have left super-abundant evidence of humanity's astonishing capacity for creativity.

Think of the various alphabets, forms of writing, number systems, calendars, mathematics, books, museums, and libraries. Think of the domestication of hundreds of plants for food. Think of wine, oil and herbal medicines. Think of megaliths and monuments built to study the seasonal return of moon, sun and stars. Think of paper made from papyrus pulp, pottery from clay, textiles from plants.

Think of metallurgy and the hundreds of uses for metals. Think of weights and measures to establish standards in barter, trade and commerce. Think of the astrolobe and telescope for navigation and to study the night sky. Think of codes of law and religious commandments. Think of windmills, watermills, aqueducts, bridges, and roads. Think of the creativity of architects and engineers. Think of the microscope and the numerous discoveries and cures for diseases. Think of music, art, literature and philosophy. Think of the theories of man, nature and the universe. In this limited number of examples, we have already sufficient evidence of humanity's inborn and evolved morphogenetic creativity.

Yet the aim of *The Creative Conscience* is to motivate you, the reader, to draw on your own morphogenetic intelligence to create a life philosophy of your own. Only you can decide what you are going to do with the rest of your life. You are in possession of innate intelligence, indwelling creativity, and a symbiotic conscience that emanated from the depths of timeless nature.

Human Conscience and Creativity

Since the body and the body-mind are both symbioses, we have strong circumstantial evidence that the human conscience, born out of animate nature, probably is symbiotic in function and purpose.

Moreover, history and anthropology provide abundant ancillary evidence that man evolved successive symbiotic societies and civilizations (e.g., the Incas, Mayans, Aztecs, the ancient Egyptians and Greeks) . In fact, generally, a civilization emerges from a culture with commonly shared customs, traditions and beliefs about existence and nature. Its artifacts, symbolic archeological structures and human bones are often all that remain.

Today sciences and humanities worldwide also provide evidence of humankind's inherently creative and conscientious nature. For instance, by their meticulous study of life forms and the earth's environment, the natural sciences identify the processes inherent in all life. On the other hand, through their sensuous, impassioned, imaginative and moral creations, the humanities display the two dominant characteristics of the human psyche. All in all, through creativity and conscience, the psyche-mind seeks to ascertain the eventual sense of life. This search for meaning is the ultimate trait of our nature-born conscience, which evaluates humankind's experiments with life's experiences.

As odd as it may at first sound, the symbiotic conscience may have been initiated in the earliest stages of our existence. To survive, hominids had to develop skill in camouflage detection, which meant the ability to distinguish a form hidden in its surroundings. Eventually, this practice in depth perception led to humankind's capacity to detect truth from falsehood. Hence the symbiotic conscience evolved judgement and understanding. In addition to the ability to define specific forms in the environment, we may gradually have developed the symbiotic skill of detecting how things fit together truly and naturally.

Yet symbiosis does not act in isolation. Symbiosis is always involved with morphogenesis and metamorphosis. What morphogenesis initiates in nature, symbiosis slowly but surely brings to stages of consummation until the most complete stage of change takes place. The mutual purpose of morphogenesis and symbiosis is to create life forms able to survive. These processes in past, present and future meld as an invisible-visible lifeline in time.

Applied to humankind, individuals need to synthesize who and what they were in the past with the persons they are now and with who and what they expect to be in the future.

The rather obvious lesson for us as individuals and as societies is that we need to learn to interact so as to benefit each other materially, mentally, and humanly. That way we will survive as symbiotic nature has taught us over the generations.

We have seen elsewhere that symbiosis in nature is the means by which morphogenetic excesses are kept in check. Apparently in the human being, the symbiotic conscience would have, more or less, the same function. In other words, though we know that in woman symbiosis expresses her compassion and warmhearted understanding, yet her symbiosis demonstrates an obverse side when she disciplines her child for its excesses.

The ancient Greeks were very familiar with the concept of a conscience that is symbiotic. True justice is symbiotic or should be. When an otherwise admirable individual displays rashness, arrogance, disregard for the feelings or rights of others (hybris), they are severely punished for it. Ancient Greek mythology and tragedy were laden with stories of heroes whose tragic flaw of excess brought on their fate and death. That is the reason Sophocles, Euripides, and Aeschylus actually commended *sōphrosynē* or moderation and temperance in life.

Thus the symbiotic conscience can also be a disiplined conscience. To carry through in life as an adult, one needs self-discipline. If one is responsible for others and disregards their needs, he himself should be disciplined for his neglect, indifference, and irresponsibility.

Moreover, it is through discipline that one achieves great skill in a sport, vocation, art, trade, business, profession or any responsible walk of life. This kind of discipline measures what it takes to succeed, but not necessarily against competition. It requires the individual to assess honestly and thoroughly what it will take him or her to reach the top. That means symbiotic knowledge of the field and keen appraisal of the skill required. It also means total commitment to one's goal. That is conscience in the highest degree of self-trust and self-knowledge. Commitment means the determination to surpass one's past achievements. Many mouth the promise. Very few have enough character to carry it out.

The Archetypal Destinies of Man and Woman

Is it possible that morphogenesis and symbiosis are expressed in different ways in men and women? For instance, in the male, morphogenesis would manifest itself as primarily masculine: in its drive to explore and penetrate the unknown, in its impulse to experiment, and in its daring creativity. (Masculine achievement would be expressed by architecture, innovative engineering, scientific research, and the Star Trek theme: "To go where no man has one before.")

On the other hand, symbiosis in nature shows itself in the design of plants and animals, in organs with purposeful functions, the vertebrate structure in mammals, and in the biosystems of offense and defense in all living creatures.

In the human male, symbiosis could express itself through willpower and anger with the dishonest and dishonorable. In addition, it could be manifest in the effort to design and perfect work and in the sustained endeavor in planning and completing long-term goals. On the other hand, the male's strict conscience and sense of absolute justice are counter-balanced by heroic self-sacrifice, bravery in rescuing others, compassion for the weak, needy and helpless.

The female is primarily symbiotic in instinctively seeking to overcome differences, to heal misunderstandings, and to clear up miscommunications. She strives to understand different versions of the truth as when arguments break out among her children. Yet the female is essentially morphogenetic in the creative power of her womb, in the matrix of meanings she sees in life, and in her emotions and intuitions that teach her what her womanly destiny should be.

Woman's morphogenetic/symbiotic nature is a guiding, mentoring force. It acts to ensure the survival of the newborn and to make sure her offspring will become as complete and perfect a person as possible. Seen in this light, the famous version of evolution as a "struggle for survival" is essentially a male animal view. This masculine interpretation of nature needs to be corrected, modified and complemented. The female's significance and importance must not be forgotten. By her role as mother, she has humanized us in our evolution. Any third millennium theory of evolution must include the feminine side of the story.

Nature's Influence on Human Character and Personality

Morphogenesis and symbiosis apparently have a direct influence on masculine character and feminine personality.

Generally speaking, masculine character seems predominantly to derive from morphogenesis whereas feminine personality derives from symbiosis. Evidence for this assertion comes from the ways the opposite sexes traditionally see each other.

The masculine is thought to express itself primarily by a range of character traits. (1) He is forceful, decisive, determined, dependable, protective. (2) Man is the adventurer, explorer, nomad, the hunter. (3) He is creative and constructive, the maker of artifacts, tools and weapons. (4) He has insatiable curiosity about the world and how it fits together. (5) On the adverse side, he tends to be aggressive and warlike. His great fault is overweening pride. (6) Yet his great merits are as father, brother, husband, friend, lover.

The feminine is thought to express itself in a range of personality traits. (1) She can be indecisive, deceptive, and vain. Yet generally, she is sympathetic and compassionate. (2) She shows skill in negotiation and in bypassing senseless confrontation. (3) Usually, she is a good neighbor, homemaker, keeper of the family unity and integrity. (4) She knows how to cook, garden, mend, and do handicrafts. (5) She admires cleanliness and order. (6) When women love, it is often sincere, wholehearted, and of the soul. They do tend us from cradle to grave. (7) Generally, women are guided by temperance, moderation, and common sense. (8) They practice sensitive, sensible wisdom in the performance of their womanly duties and responsibilities. (9) She is the mother, sister, wife, friend and beloved.

Of course, a skeptical reader may object to these viewpoints as simple stereotypes. Nevertheless, these traditional views do correspond closely to the ways men and women see each other.

On the other hand, it is obvious that no man is pure masculinity and no woman is unadulterated femininity. As the Swiss psychologist C.G. Jung (1875–1961) explained in his analytical psychology, every man may be dominated by his masculine animus, and every woman by her feminine anima. However, consciousness of being one sex is usually complemented subconsciously by awareness of a somewhat opposite temperament.

There are two reasons for this cognizance. For one thing our actual sex is determined in the months just prior to our birth. Secondly, from our first breath of life, our role models are father and mother. Our memory of them influences the rest of our lives, in mental outlook, behavior and conduct.

For instance, like symbiosis the animus can express itself at the conscious and unconscious levels. In males, both symbiosis and animus may predominate. By reason of their continuous contact with the outside world, competition with other males and their sense of responsibility for the family's survival, symbiosis in the male animus takes on distinct characteristics. They learn to plan and execute tasks and long-term work. If their symbiotic nature enables them to design and reason, their animus animates or activates the coordination to construct shelter, build roads, arrange irrigation systems, and the like. Similarly their effective action produces worldwide networks of communication. Guided by the animus, whatever they do has purpose and what they build is structured to withstand time. Men have often the need to achieve craftsmanlike accomplishments in which they can take lifetime pride.

In women, their basically feminine, morphogenetic nature may be energized and focused by their subconscious animus or masculine side. For instance, it may be responsible for her curiosity, creativity, and experimentation. Women are also endowed with originality, capable of inventing all kinds of things needed in domestic and practical life. (In traditional societies, their herb-gathering and knowledge enabled them to alleviate pain and heal wounds as well as to mitigate and cure sicknesses.)

In the twentieth century, their patience and keenness gave us major breakthroughs in medicine and in microbiology, and generally, they make excellent researchers in all fields of science. Women make superior teachers, nurses, and medical doctors. Moreover, as regards human rights, they are known for their conscientious commitment to mutual understanding among individuals, ethnic groups, and different races.

So biologically speaking, man and woman share nature's morphogenesis and symbiosis. Psychologically speaking, they share animus and anima to provide each other with character and personality. Both man and woman are gifted with creativity and conscience, with originality and a deep instinct that we share a mutual purpose.

If the male psyche teaches us the worth of determination, steadfastness and endurance, the female psyche teaches us the value of caring, compassion and humaneness. If, after all, morphogenesis and symbiosis are the true source of the evolution of animate nature, the dialectic between male and female, animus and anima may be the ultimate reason for human evolution.

In any event, it is abundantly clear that nature meant man and woman to support and strengthen each other. The true man and the true woman together can fulfill each other's lives. Together, they can stand unafraid of whatever fate or destiny may bring humankind.

Paul Diel's Reinterpretation of Psyche

The French psychologist Paul Diel provides us with a different insight into psyche. He describes the subconscious and supraconscious in the context of evolution.

The basic premise of Paul Diel's *Psychology of Motivation* is that soma and psyche form a biogenetic unity. As the human species evolved, soma and psyche evolved concurrently. If there is present in psyche a function we may name subconscious, as defined by Freud, Paul Diel maintains that the pan-sexual interpretation of psyche ignores the true cause of mental illness, which occurs when human vanity calls forth senseless, exalted desires bound to be frustrated by the requirements of the real world. More importantly, the human psyche manifests a supraconscience as foreshadowed in ancient mythologies and in man's notions of the Divinity.

In such scriptures as the Old and New Testament, Diel contends that the eternal truths and intimations of immortality are mythical expressions of humanity's awakening awareness that man is endowed with a supraconscious and that life has a biological sense. For Paul Diel such myths represent the psychological prescience of ancient peoples which can be developed into a true science when such myths are adequately interpreted in the context of contemporary depth psychology.

Hence Diel's psychology depicts how the demons and powers of darkness in myth are psychic manifestations welling up from the subconscious whereas the gods and powers of light are signs of man's supraconscious. It is the intimation of this higher mental power which motivates man to subordinate and harmonize all his material, sexual, and spiritual desires to his essential desire, which is to find the sense and direction of human life. (Strauch, 158)[1]

Throughout *Psychologie de la Motivation*, Diel describes the intricate interplay of opposing forces. The subconscious tends to distintegrate the personality whereas the supraconscious undertakes to transform, heal, and integrate the personality into a definite identity with a purposeful destiny. (Strauch, 158)

The Subconscious

Diel distinguished a number of characteristics which define the subconscious. It begins with committing the vital sin, which comes from exalting desires that are false and from failing to satisfy the vital need of the essential desire. (That vital need is to realize the sense of life.) Moreover, to chase multiple desires is to fail to pursue life's essential goal. (Diel, 80–81)[2]

The Supraconscience

In contrast to the lifetime punishment we experience in chasing senseless, multiple desires, the pursuit of the essential desire rewards us. Through that pursuit, suffering can be sublimated and spiritualized. Rather than being misled by fatal choices, when we obey the voice of the supraconscious, we are able to make sound choices and realize a meaningful destiny. The essential desire centers and unifies the conscious self into the conscience aware of its moral responsibilities. (Diel, 182)

The sentiment of having satisfied life and being satisfied with life is the essential success. Essential success is strikingingly defined: "The ideal combat is not to want to surpass others, but to surpass oneself in the evolutionary sense." (Diel 162) "The authority of the essential desire is most clearly seen in its power to concentrate the psychic forces. Through the sound education of the essential desire, the individual is rewarded with life's essential success." (Diel, 76–77)

Morphogenesis and Symbiosis in the Context of Diel's Psychology

Morphogenesis means creativity, ontogenesis and evolution, whereas symbiosis means somatic-psychic integration, which ultimately means both conscience and supraconscience.

Naturally, human curiosity explores a variety of ideas, which, despite Diel's criticism of pursuing multiple desires, may not be undesirable. In its own right, such exploration may be a form of experimenting with life's experiences to discover the essential desire of one's life.

On the other hand, the purpose of the creative conscience is to integrate life's experiences into one essential meaning. Its role is to guide, mentor, and perfect our moral evolution.

The Consequences of Co-Evolution

The reader will recognize certain theses in *The Creative Conscience* that are reflected in the basic assumption of Diel's philosophy.

1. His psychology is based on the assumption and conviction that soma and psyche co-evolved. In my own presentation, hominids co-evolved with the cultures and civilizations they themselves created.

2. Further, the book's thesis maintains that the human body is a biogenetic entity by reason of its evolution and integration. The body itself is the result of the co-evolution of all its lifesupport systems.

3. Moreover, the brain itself must have evolved along with the body so that body and mind evolved in tandem.

4. In addition, the distinct natures of man and woman may also have co-evolved, making their intimate relationship a dynamic factor in hominid evolution. As woman adapted to man's needs, man adjusted to woman's. Mutual understanding became necessary. This intimate harmony made family, society and culture possible even in remotest history.

5. Moreover, the mutual evolution of man as animus and woman as anima may have contributed to the humanization of our ancestors. As family, the couple likely accelerated their moral maturation by teaching each other the meaning of human responsibility. This psychic metamorphosis might be the evolutionary significance of a conjugal, connubial lifetime together.

The Supraconscience as Creative Conscience

Diel gives us an admirable description of the superconscious or supraconscience. His most convincing argument is that, in contrast to pursuing senseless, multiple desires, the pursuit of the essential desire in life will bring lifetime satisfaction. Certainly the effort to find the sense and direction of one's life strikes a chord of infinite wisdom in those who have thought about the meaning of life.

Furthermore, Diel depicts the supraconscience as transforming, healing and integrating personality. It does so to create a definite identity with a purposeful destiny. Clearly this portrayal has a marked kinship to the argument presented in *The Creative Conscience*.

On the other hand, his work is inspired by a generalized idea of the influence of the evolution of the psyche. Examples drawn from Greek mythology, the Bible, and the concept of the Divinity admirably illustrate his comprehensive descriptions of the subconscious and the supraconsciousness.

My own work is founded on scientific descriptions from biology, ecology and microbiology. The examples used have undertaken to humanize the meaning of these life sciences and to show their value in layman terms. Moreover, the intention is to encourage the individual to derive a life philosophy from the ageless facts and phenomena that created, governed and evolved the human psyche, mind and intelligence.

Metaphor and Irony as Keys to Human Destiny

Connie Barlow's (ed.) *From Gaia to Selfish Genes: Selected Writings in the Life Sciences* (1991) inspired a number of insights into evolution as the integration of forms of life. Her thesis helped me realize the universal significance of metaphor and irony. Indeed, they reveal the morphogenetic/symbiotic essence of the human mind itself.

Metaphor has the ability to transcend morphogenetically the logic of "either/or" or of "sound" reasoning. At the same time, metaphor illustrates the capacity to integrate what appears distinct and disparate. (Symbiosis, as biological integration *within* life forms and as ecological bond between species, establishes mutual associations, relations and meanings to life processes.) By extension, through images and intimations, metaphor used in poetry and holy scripture unveils life's most sublime and significant experiences.

Belief in the invisible, concealed by the visible, characterizes not only the superstitious but actually defines the essence of religious faith and scientific research. This surprising, unexpected connection between what we usually call the "irrational" and the "accepted rational" should give us pause for further thought. Obviously, the sciences make little or no effort to examine this twilight zone of human perception.

Then there is the use of irony to express our viewpoint on events, circumstances and conditions of the world. Clearly, the sciences make no effort to account for irony, and yet nature, existence, life and humanity are intertwined symbiotically with an infinite spider web of ironies that seems to reach from the earth to the outer reaches of our universe.

Any individual who reaches the age of maturity becomes aware of the incongruities, paradoxes, and enigmas inherent in nature, in existence, life and human nature. Much is beyond man's logic and understanding. It is the use of irony that reveals the human mind aware of self in the world and existence. By the astute use of irony, we not only comment on the incomprehensibles around us and the ineffables in us. Marvelously, the skepticism of irony provides a necessary counterpoint to the zeal, nobility, and "ecstasy" of life's metaphoric truths.

In a way, irony is fable whereas metaphor is fairy tale. Irony is the bitter wisdom of the Old Testament whereas metaphor is the faith, hope and charity of the New Testament. Moses saw the irony of the lascivious, greedy, immoral Israeli as God's Chosen People and realized they needed to be reminded of the immense power in the universe. Thus the Ten Commandments taught them fear and respect for that power. The aim was to return them to self-discipline.

On the other hand, Jesus of Nazareth saw through the weaknesses, selfishness, duplicity and intolerance of humankind. By using metaphors, parables, and symbolic stories, he awakened the hope that his followers could overcome their fear of hunger, thirst, sickness and death. Jesus illustrated the truth that the human heart contains all the knowledge and wisdom needed to live a life worthy of God's love.

Hence both irony and metaphor are the means by which the human mind acknowledges the bitter realities and desperate hopes of humble humanity.

The power of irony and metaphor arises from the bio-psychological fact that body and mind are not only born of animate nature, but also that they are the foci of concentration of our morphogenetic/symbiotic nature.[3]

By a kind of dialogue and deliberation between creative morphogenesis and our symbiotic conscience, we are able to penetrate appearances and perceive actual reality. Thus our natural conscience is able to discern the enduring truths beyond the ephemeral deceptions of life.

Let us briefly recall the significance of this imaginative use of language. Metaphor is a means of analogy. Between things thought to be distinct, one discovers certain similarities in quality or essence.

Metaphor is primarily a mode of intensification and secondarily of similitude. It intensifies mood or state of mind by its intentional ambiguity. When we recall that language used for its positive ambiguity implies widening spheres of significance, we see how metaphor is metaphor by reason of the multiple expectations and inferences it evokes. Metaphor aims primarily at arousing wonder at a hidden relationship, mystery or truth.

Moreover, a metaphor is like an enigma in prompting us to solve the mystery of the many associations it calls forth in us. In a sense it is a riddle, a spell, an oracle all at once, and it shares these qualities with irony. Like irony, metaphor widens the implications of things, but unlike irony which detects the fact that an ugly reality often contradicts a comely or benign appearance, metaphor sees a spiritual reality beneath the deceptively plain appearance of mundane things. In other words, in contrast to the dualistic vision of things inherent in irony, metaphor pierces the illusion of duality of self against existence to uncover the underlying (symbiotic) oneness of life. (Strauch, *A Philosophy of Literary Criticism*, 1974, 119–121)

Metaphor and irony are the two most pervasive world views in world literature. Over and again we encounter authors with a predominantly metaphoric vision or ironic perception. The metaphor is hopeful, spiritual, even mystical in its faith that there is a power of life guiding us to a noble, meaningful destiny. On the other hand, irony is realistic, skeptical, sometimes cynical in its conviction that life is tragic and that mankind's excesses can only lead to tragedy. The ironic vision sees mankind as fated. Yet both metaphor and irony teach us to seek out a reliable reality beneath the world's deceptive appearances or to go in quest of a timeless verity beyond the visible illusions of this brief life.

In sum, by using metaphor and irony, we express the true nature of our evolution as the mind defines the meaning of human life.

In addition to the evidence provided by language and literature, we find that philosophy has used two forms of dialectical reasoning parallel to the literary use of metaphor and irony, and this use offers further corroboration that these world views are universal as are the processes morphogenesis and symbiosis and human creativity and conscience.

For instance, there is the work of the German philosopher G.W.F. Hegel (1770–1831). His theory of the dialectic active throughout history may be regarded as fundamentally metaphoric in Weltanschauung. His is a positive dialectic in the sense that it seeks to overcome misunderstanding, disagreement, confrontation and conflict.

Developed through stages of thesis and antithesis, the Hegelian process finds resolution in synthesis. In Hegel, the impasse between opposing principles will hopefully lead the disputants to realize that a higher truth may remove the antimony between them. In other words, when the synthesis is reached, a greater understanding resolves their differences. By being optimistic, Hegel's dialectic is both positive and metaphoric in inclination. Both dialectic and metaphor are spiritual in nature.

In sum, because the purpose of his dialectic is synthesis , inclusion and integration, his form of reasoning is an example of our symbiotic conscience in philosophy.

By contrast, there is another, opposite way of reasoning also characteristic of human conscience. Ever since antiquity, Greek drama used irony to illustrate the tragic view that human life is fated. The ancient Greek philosopher Socrates (ca 470–399 B.C.E.) was the central figure in Plato's Dialogues. To counter the dishonest, specious arguments of the Sophists, he developed a methodical, ironic form of reasoning. The dialogues *Apology*, *Crito* and *Phaedo* dealt with his trial and imprisonment for teaching the youth to test the unquestioned assumptions of their elders.

When others called him a wise man, he declared that his only wisdom was the knowledge he was ignorant. So his lifetime preoccupation was to find the truth and to help others find it as well. His philosophy was. "The unexamined life is not worth living." He conceived the highest state of mind to be arete, the moral knowledge that guides the individual to decide the best way to find the truth and to live life.

Socrates provided the philosophical foundation to Western civilization both by his honesty and his systematic pursuit of the truth. His method of questioning cut through vague understanding, misconceptions, and pseudo-knowledge by examining every cherished assumption for its soundness, accuracy and rightness.

Socrates' search for irrefutable truth was basically a negative dialectic by its process of analysis, by its exclusion of the unsound, and the elimination of the untenable.

Hence, between the positive dialectic of Hegel and the negative dialectic of Socrates we recognize two archetypal ways of visualizing existence. These polar points of view seem to corroborate how the human conscience has evolved. Through their example, it would appear that our morphogenetic/symbiotic conscience is itself a dialectic. Perhaps the dialogue between human intelligence and life may yet evolve humankind into sapient Homo sapiens.

12

THE PLAN AND PURPOSE OF NATURE

Introduction

As we have seen, Nature created all life. It designed life forms that could endure long enough to reproduce themselves. Moreover, in response to the changing conditions of earthly existence, nature enabled creatures to evolve so as to continue their life line across hundreds of thousands of years. To accomplish this, Nature developed and pursued a plan and a purpose. That is the only sound explanation for the endless creation of the infinite variety of the earth's life forms.

Against this undeniable, universal evidence, all attempts at explaining animate nature as illustrating the doctrine of mechanism or biological automatism are patently false. Moreover, causal explanations of the human mind are equally useless, especially if one tries to explain why ideas emanate from our imagination. Nor can scientific materialism account for the way ideals can empower the individual and transform entire populations to pursue a life of moral purpose and meaning.

Furthermore, no biological determinism can account for the evolutionary power of nature's Creative Conscience , which made possible the mutation of human intelligence. Nor can any predestination account for the capacity of humankind to create cultures and civilizations. As we co-evolved, we learned to create and trust nature to guide our conscience. Our human creativity and the evolvement of our natural conscience influenced our evolution from primate to

Homo sapiens. In the deep nature of humanity is manifest Nature's Creative Conscience.

Hierarchy in Nature

Early man understood nature as a hierarchy among insects, plants, reptiles, birds and animals. (Aesop's fables illustrate our early understanding of our fellow creatures.) In traditional societies there emerged social strata with chief or priest at the acme. There followed elders, fathers, mothers, and finally children from the eldest to the youngest. Other hierarchies probably existed according to hunting, farming, medical or other survival skills. According to one's utility and intelligence, the society held the individual in esteem.

In the human body itself, there appears to exist a hierarchy similar in purpose to that in open nature and traditional society. The purpose of all the activity of our cells, tissues, circulation, respiration, metabolism, senses and the nervous system is to maintain our health and survive. In other words, the body's life-support systems are to keep the heart and brain alive, the acme of our being.

In fact, an individual is declared dead only when the brain dies. So all this intricate morphogenesis and consolidated symbiosis of systems is to sustain the soundness of the mind and to keep the brain alive. Since this fact is self-evident, the human brain must be the highest order of morphogenesis/symbiosis on earth.

Not only is its protection, development and education of the greatest importance. The realization of its creative and cognitive powers seems the very purpose of Nature itself.

In any event, humankind needs to grasp more fully how and why the brain does what it does. Illustrating the mind's capacities is not a matter of drawing facile analogies between the human mind and, for instance, the computer, gasolene engine, television set, cellular phone or any other electronic device. These are merely ingenious toys compared to the human mind that invented them.

Rather, the morphogenetic mind is a symbiosis of life-giving networks, each with its own lesson in logic and purposeful activity for cognitive scientists and philosophers. Above all, the mind's total, holistic activity is committed to ultimate meanings.

Thus throughout physical and cerebral nature, there appear hierarchies which serve necessary functions aimed not only at perpetuating life but also at serving a higher purpose in humanity.

One day, morphogenesis and symbiosis will be acknowledged to be the bio-processes largely responsible for humanity's highest intellectual achievements and for the ageless wisdom of cultures the world over.

Survival and the Acceleration of Intelligence

In the beginning of life on earth, there were earthquakes, volcanic eruptions, typhoons, other natural disasters and chaos. In order to survive, the simplest life forms had to develop senses and a degree of intelligence. Eventually, they developed multiple defences, varied strategies of survival and life processes that enabled them to evolve. Chief among these were morphogenesis and symbiosis,.

So the original chaotic conditions of earthly existence engendered and integrated two kinds of consciousness, one with an indeterminate, creative potential. The other, complementary in nature, became responsible for defensive systems, cautious responses and self-restraint. Ultimately, these corollary modes of consciousness became a form of natural conscience in humankind.

Gradually, these faculties grew capable of rapidly judging situations, challenges, and needs in the light of past experiences. Memory led to thought. In the long run, these proto-forms of consciousness and conscience developed the capacity to foresee problems of the immediate and distant future. Eventually, a form of brain was needed to coordinate every cell, tissue, nerve and system in the body somatically, sentiently, noetically, and creatively from split-second decisions to decisions pursuing a lifetime purpose.

It seems possible that survivors would have survived through the accelerated development of their morphogenetic creativity to meet repeated catastrophic changes. Correspondingly, as a survivor learned to utilize the knowledge gained from experience, its symbiosis would accelerate its potentials. As such, symbiosis functioned to coalesce, consolidate, and integrate survival information. At a given point in their interaction, symbiosis exerts control over the morphogenetic process, restrains its excesses, guides it, as it were, educates it. Although, symbiosis becomes the organism's survival consciousness or conscience, in time of urgency, morphogenesis and symbiosis can, effectively accelerate each other's efficiency and effectiveness to meet the organism's survival needs. This capacity for acceleration may account for saltations in mammalia, primates, and hominidae. A saltation can indicate the origin of a new species such that a single evolutionary step is made, and it may account for a major mutation. Hence it may be possible that the interaction of morphogenesis and symbiosis as processes, developed by human intelligence, actually contributed to the acceleration of the change in size of the human skull.

In hominids it might account for the successively larger brain of man/apes, Homo erectus, Cro-Magnon man, and the Homo sapiens of today. (Anthropologists maintain that the extinct Neanderthal had the largest cranium.)

However, the concern here is less with the size of the cranium than in what it contains. Since nuclei came later in evolution,[1] that would seem evidence of a mode of evolved morphogenetic/symbiotic integration by providing a center, a hub, or a source of intelligent instructions for the cell's cytoplasm, etc. From that remote single cell beginning in the evolution of living matter, there seems to have emerged the intelligence of animate Nature. The cell's capacity for morphogenetic transformation and symbiotic integration, capable of self-reproduction and consolidation with others, may be the clue to the growth of the brain within the confines of a cranium.

Rather than gauge the size for successive stages of hominid evolution, its makes more sense to consider the intricate convolutions of the brain. (It is said that the greater the intelligent use of the brain, the more evidence of convolutions.)

It is the contention of my theory that the brain is a cluster of neurons in intricate symbiotic combinations which facilitate the quantum leaps of morphogenetic imagination of the human mind. Not only are evolutionary saltations due to morphogenetic/symbiotic interaction but also are due to our quantum leaps in perception and conception. That capacity of the human mind is also evidence of our accelerated evolution.

Intelligence as the Consequence of Co-evolution

It is probable that human intelligence is the consequence of the co-evolution of three correlative evolvements: (1) our biological evolution; (2) our succeeding cultures (religions, humanities); and (3) our civilizations (their rational and legal systems as well as their inventions, skills and innovations).

Understanding human nature is important because our evolution has been distinct. Through the co-evolution of our biogenetic being and the cultures we created, human intelligence emulated the primal evolution of Nature itself.

We must not fail to keep our conscience alive, nor fail to continue our creative pursuits. To neglect our conscience would be to fail to consolidate and assimilate what life teaches us about ourself and our humanity. To neglect our creativity would be to fail to explore our talents, aptitudes, curiosity, personality and the joy of discovering who we really are through self-expression. To not cultivate these propensities of human nature would be to neglect the promise of a fuller, richer destiny in keeping with the explorative, creative history of humanity.

To fail to understand the power of the creative conscience in our personal lives would be to remain stillborn psychologically and spiritually. Or we would remain aborted human beings in that we failed to nurture the emotional, artistic, and intelligent needs of ourselves and others.

The key to our humanity it its arts, music, games, sports, literature and life philosophies. Moreover, the expression of our ecstatic nature in myth, religion, mysticism and the search for life's meaning evolved our emotional intelligence. Our true humanity takes place when we become trustworthy, helpful, charitable, kind, and morally responsible.

Morphogenetic Human Nature

All life forms manifest some degree of self-direction. This fact of nature should make it clear that each individual has an inherent identity, and he identifies himself by living out a special destiny. In humankind, one's destiny is made manifest through the development of special talents and the cultivation of intelligence.

In the appearance of a genuine genius, his or her self-realization illustrates more than an exceptional individual's destiny. At the very least, each genius provides us with lessons in self-education and in self-exploration that should guide all of us to seek to create a special life of our own. Yet human genius is not solely evidence of human evolution. Rather, it reveals the heritage Mother Nature has bequeathed humankind out of the inexhaustible plenty of her own Infinite Genius.

The human mind continually spawns ideas, dreams, keen perceptions and original concepts. The brain has evolved by the continuous exercise of ingenuity pursuing the fulfillment of some worthwhile purpose. Though life may impose limits on us, the creative imagination will seek a way out. However opposite the conditions of existence, however senseless may seem the times, the mind has the power of visualizing a better future. Endowed by nature with the capacity to imagine, the mind will always find ways to soar beyond space and time as if it were born to be free.

Just as our bio-nature was energized by our morphogenetic creativity and guided by our symbiotic conscience, so too our cultures and civilizations co-evolved with our biological evolution. To be sure, they actually came about by stages of development of human intelligence. They activated and actualized different stages through our own mental development.

History seems to provide abundant evidence that such a supra-evolution has actually taken place for humankind across the millennia of archeological and recorded history. This evidence requires a qualification, however.

While it is easy for skeptics or "realists" to remind us of the decline and fall of civilizations and the mysterious disappearance of entire populations as in Mesoamerica, yet the converse appears to offer a truer understanding of humanity's history. After all, for generation upon generation and for age upon age certain vestiges, ideas, skills and symbols have endured and become refined. Indeed, artifacts have enabled other races in later ages to establish richer cultures and superior civilizations. These should remind us of how human intelligence from the past has helped us survive, despite the most awful events that have happened before.

Hence the history of the world is also a record of human achievement—unimaginable, unbelievable, incredible in view of our prehistoric beginnings. Yet we are still here thanks to those who came before us and who remain in memory to guide us.

The Need for Creativity and Conscience

Twentieth century thinking shifted away from the emphasis on rational and causal explanations of the human mind to the investigation of its inner, psychic dynamics. As has been demonstrated, morphogenesis and symbiosis generate and initiate all our bio-functions. Consequently, they must similarly affect the human psyche and mind. We need to bear in mind that these two processes are active in us biologically, emotionally, mentally, creatively and in the conscience born out of hominid evolution.

Nature's Processes Account for Our Mental Evolution

We have defined morphogenesis as our instinctive curiosity and exploration of our outer and inner environment. As such, it contributes to our genesis as an individual and the evolution of our species. By its curiosity as to what we are, morphogenesis accounts in part for our habit of introspection in order to ferret out the unhealthy from the healthy. Thereby it enhances our potential for survival.

Thus morphogenesis has two main functions. First, it is responsible for adapting us to the world outside our body. Second, its intrinsic activity accounts for the evolution of our life support systems in response to environment challenges.

In view of the fact that our memory and imagination are activated by image, metaphor, and symbol, the psyche itself manifests morphogenesis. Such imagery seems to germinate, grow, and flow into symbiotic symbols which have both meaning and purpose.

If morphogenesis and symbiosis are, in fact, the two ubiquitous processes actuating and actualizing life in all its forms, their interaction accounts for our mental evolution over millions of years.

As observed before, by and large, symbiosis guides, controls, and moderates morphogenesis to prevent its tendency to creative excess, as in the ancient Greek concept of hybris. However, by the same token, morphogenesis exerts its "rights" against excessive control, restraint, or suppression which symbiosis might, on occasion, be seen to exert. Nevertheless, in general, symbiosis acts periodically as *sōphrosynē* (moderation, temperance) in order to maintain the life form in a healthful homeostasis, still with the capacity of further self-transformation.

Our past and present as a species reveals us to be creative symbionts. Our survival thus far urges us to continue our symbiotic destiny. Yet our innermost nature spurs us to continue to evolve physically, genetically, and mentally. After all, evolution is both an unfolding of potentialities and the creation of new potentialities. Between old and new forms of life, between old and new perceptions and concepts, the dialectic between old and new strategies of life, old and new life philosophies continues. Thus it would appear that morphogenesis and symbiosis interact homeostatically to bring about periods of stability and periods of creative evolution.

Homo sapiens may turn out to be the species that instinctively needs to discover meaning, so that our sometimes mysterious and baffling passage through life not simply end in a meaningless zero.

Through time, humanity has lived either morphogenetic or symbiotic destinies, i.e. we have lived either creatively or conscientiously. The individual lives through successive spheres of time. These times become spheres of significance when the individual, via intelligence and intuition, learns to perceive that each stage of life has a purpose. Thus each sphere of time teaches the individual there is a certain meaning to life, and through this insight, he or she can grasp the message life brings.

Given the fact that we emerged out of the obscurity of time, we have learned that we have a measure of control over the time we do have to live. If that impression is not mere illusion, our intimation that life has a purpose seems an invitation to realize a destiny as individuals and as a special species on this planet.

How Nature Teaches Us to Survive

Humankind's symbiotic nature explains much about our religions, societies, cultures and the psyche-mind. That symbiotic nature obliges Homo sapiens to come to terms with four existential needs, if we are to create a world where families can live without fear that one day they will be destroyed by man's abomination of man.

Four Symbiotic Principles for Survival

1. The individual should come to terms with his physical and mental needs so as to achieve a sound mind in a sound body.

2. The individual should seek to integrate what he has learned so as to evolve a general, generic knowledge.

3. Humankind need to explore more meaningfully the ways and means of achieving international reciprocity, which means a greater sharing of the world's resources—mineral, agricultural, medical, educational. It means universal mutual aid.

4. Human societies need laws to be more symbiotically just where punishment and crime are in sensible proportion. Moreover, incarceration and ineffectual rehabilitation programs are not the answer. (Can the criminal be educated to use creatively his general intelligence, aptitudes or talents?) For prisoners to be liberated, a condition should be set that they achieve a certain level of literacy, of education, or of vocational skills. In addition, a patient effort to morally reeducate them should be made so that they can live as decent, responsible citizens, once released.

If symbiosis is a universal law in nature and in ourselves, humankind should be taught that benign, life-giving principles and commandments should be heeded and obeyed.

When humankind become genuinely symbiotic beings in mind, spirit and humaneness, then we will be ready to begin our journey to the next level of our humanity.

The Symbiotic Conscience

Of all the life support systems in the human body, it may be the immune system which seems most directly linked to the body's memory of past events. In a sense, we may designate that memory as part of our symbiotic conscience. At this organic level of survival, morphogenesis and symbiosis together effect a defense against microbial invaders and initiate an offensive to overwhelm the enemy. In this respect, our symbiotic conscience has the responsibility to promote our continued survival. This evidence alone would make it appear that the human conscience is a result of the evolution of our systems of self-preservation through morphogenetic resourcefulness and symbiotic self-command.

The symbiotic conscience of our species teaches us to incorporate what is good, healthful, and life engendering and to eliminate what is bad, contagious, and life-threatening. Consequently, our natural conscience has the intelligence to know right from wrong, the true from the false, the moral from the immoral.

The bio-basis for human conscience enables us to reach decisions that make sense and make decisions that reject the senseless; to act according to the true and to avoid the untrue; to believe in the benign and to be wary of the malign. Evolution endowed us with the intelligence to survive existence, and our survival as human beings required that we evolve the symbiotic conscience.

Consequences of Human Evolution

In previous chapters we retraced the universal manifestation of morphogenesis and symbiosis in bacteria, cells, insects, plants, and animals. In humankind the body's integrated life support systems and the body-mind itself provide evidence of the same fundamental bio-processes at work. The evidence corroborates that we are endowed with morphogenetic creativity and a symbiotic conscience.

This fact has a number of consequences. For instance, it should influence the way we acquire knowledge. Apparently morphogenetic intelligence needs to be challenged. That means our inherent creativity should be exercised in both traditional and innovative ways. In other words, the creativity not only of the young but of all ages should be challenged and rewarded. Such activity will teach individuals about their capacity for originality and thus enhance their self-knowledge and self-esteem.

Similarly our symbiotic intelligence needs to be fostered. This may be encouraged by establishing a goal for knowledge- seeking. Its aims should be the holistic integration of what we learn. That would mean that knowledge is to be sought in tentative wholes. This process would be like the growth and the consolidation of tissue, bone, organs, and life-support system of our own body.

Or we might study society for its morphogenetic properties and symbiotic principles. We would need agencies or procedures to facilitate positive social change, to improve living conditions, and to ensure employment opportunities. Virtually free education can be made available to all walks of life and all levels of society, particularly for the unemployed and the homeless.

In a way, we need to consider society as a dynamic homeostasis such that its multiple activities, its citizens, and its distinct ethnic groups work together to achieve mutual goals and to ensure that mutual aid is shared throughout the society. Thus cooperation becomes coordination and integration of our best efforts to ensure security, freedom and education for all our citizens. In this way, many will be inspired to pursue life-worthy goals.

Another consequence of the morphogenetic/symbiotic understanding of human nature is to focus on the advancement of human intelligence. The goal would be to encourage the individual's inherent curiosity and creativity. On the other hand, the need for emotional fulfillment could be addressed by the encouragement of like-minded people joining together for mutually enlightening activities. To be sure, the aim of any society is to satisfy humankind's instinct for moral justice and to educate further the compassionate conscience evolved by humankind.

As to the individual's symbiotic intelligence, the individual must learn to meet the needs of body and mind. We need to come to terms with the fact that our common sense and reason have to take into account our feelings. As Blaise Pascal observed, "The heart has its reasons that reason knows nothing about." The acceptance of self and others is central to harmonizing head and heart, emotions and intelligence.

We are already aware that human history offers sundry examples of humanity's creativity and moral understanding of life in proverbs, myths, religions, customs, traditions, and the experiences of many peoples across the earth. Moreover, today's sciences and humanities are extensive realms of knowledge and creativity which indicate how human symbiotic and morphogenetic intelligence enable us to discover and create meaning from life's experiences.

These observations allow us to draw a number of meaningful conclusions.

1. Whereas nature provides evidence of the survival advantages to practicing morphogenesis and symbiosis;

2. Whereas humankind are most likely to survive by emulating nature's lessons (e.g., the mutual aid among social insects as ants, bees, wasps and termites);

3. Whereas the human body itself is a super-symbiosis of intricate, interfunctioning, self-regulating life-support systems;

4. Whereas the sciences, humanities and humankind's wisdom manifest the exceptional capacities of the human mind, it should follow that from the earliest evidence of human culture that humanity's destiny has been guided by our nature-born creativity and conscience.

Moreover,

5. Whereas we ourselves emerged from nature and have learned many of its lessons in survival;

6. Whereas our integrated, self-sustaining bodies teach us to monitor our emotions, excesses, habits, decisions and actions in order to survive, we are nature's child, body and soul;

Furthermore,

7. Whereas we evolved by cultivating our imagination and universalizing the meaning of moral conscience; *it is self-evident that human nature defines itself by its creative conscience.*

Past and present show our species to be sustained by creative intelligence and a conscience infused with a sense of justice and compassion. Hence it is probable our future history will confirm our intention to pursue a higher destiny by using our aptitudes talents, innovative skills and our intelligence. If so, humankind will continue to evolve mentally, creatively, and morally in line with our innermost nature. By conscientiously and creatively pursuing ever new knowledge and by consolidating what we learn into symbioses of understanding, experience and wisdom, humanity will surely fulfill its special destiny on earth.

On Remembering Our Own Heritage

Morphogenesis and symbiosis work together until the entire human embryo is completed and ready to be born. Thus at birth we are morphogenetic/symbiotic beings. Instinctively, we have not only experienced these life processes; somehow the newborn "knows" that throughout life it will be guided and directed by nature's creativity and wisdom. Therefore from the start, we have an intuition of the way life should be lived if we do not ignore our deeper biological heritage.

On Moral Education

We could study nature as a series of experiments, by which it evolved the conditions that guaranteed survival. In the morphogenetic sense, nature may be urging us to discover how far our intelligence can take us. However, as human beings we have already discovered that our moral evolution will not come about by intelligence alone. It simply is not enough.

Judging by the sophistry, casuistry, and dupery in many walks of life, reliance on intelligence alone does not assure us we can abide by intellectualized moral principles however noble and worthy they may be. The same reservation applies to too great a trust in conscience ruled by strictness. Intelligence devoid of conscience and conscience devoid of empathy tend to self-righteous injustice and the temptation to superimpose one's own self-assured view of life on everyone else. Insolent intelligence and excessive conscience have been the source of mutual hatreds, confrontations, and even senseless wars. So our human experience teaches us the danger of extremes.

By contrast, humankind's symbiotic conscience has shown us that empathy and compassion must temper the excesses of blind justice. These emotions are needed to moderate our meanness, intolerance, self-assurance and sense of superiority "over" others.

Indeed, humankind might aspire to moral progress if we undertook to perfect our humaneness not only in the treatment of animals but also in the care of human beings. (At times, one wonders why there is no humane society for stray and lost individuals.)

In this age when pride and blindness to consequences can trigger a war of mutual destruction, it seems time for humankind to apply morphogenetic creativity to all our old assumptions about races, religions, and political beliefs different from our own. It is time to start negotiating a moral understanding with all humanity. That would be a measure of moral progress in the evolution of humankind.

So as ancient wisdom discovered the need for moderation and compassion, so too 2,500 years later we need to listen to the lessons of human history and hear the prompting of our humane heart if we are to avoid the excesses that may destroy us and dehumanize those who survive Armageddon.

The Creative Conscience as Life Philosophy

It is important for the individual to define a life philosophy of one's own.

The philosophy can be based on the confidence that nature itself is a creative conscience and that human evolution has been due to the intricate interaction between our capacity for creativity and our awareness of the definite need for a conscience to govern our excesses and to guide human relations.

It is to be noted that conscience in nature has been made manifest by the evolution, completion and perfecting of life forms. In human terms, whatever we undertake to accomplish should be done with patient effort so as to complete and perfect the work we do.

In the process, we transform ourselves in whatever occupation, vocation, hobby, organization or calling we find ourselves. We do so by emulating the evolution omnipresent in nature and in our hominid evolution. By our contribution to the group, we ourselves evolve. In addition, those who pursue a goal, knowledge or self-education tend stage-by-stage to universalize their special intelligence and their unique creative conscience. By so doing, the individual replicates the evolution of humankind.

As a long-term commitment, pursuit of knowledge can bring the intelligent, the gifted, and conscientious the ultimate gratification that only a lifetime well-spent can give.

To be sure, the conscience evolved by hominids manifests another, fundamental side by being responsive, charitable, and compassionate. In a hostile and unforgiving environment, this form of conscience was born out of thoughtful reciprocity among humankind to ensure common survival. As an intimate sphere of cooperation and mutual aid within the larger sphere of the predatory food cycle, humanity established strategies for surviving and charitable ways of helping others survive. That inner, cooperative circle of survival not only had to become well organized, efficient and unified to fend off predators. It also gradually learned to respond to others' hunger, thirst, desperation and misery.

To be sure, groups, clans, tribes, and nations were primarily concerned with the survival of their own ways and beliefs. Yet ever since humankind began to gather together, their common aim must have been based on a sympathetic consciousness of the mutual aid all humans needed.

It is that archetypal symbiotic conscience that remains the source of hope for future humankind, the faith that we are existents and symbionts sharing one communal wish to survive and reproduce, to educate our young, and to create a world free from mistrust, suffering, sickness, malingering, duplicity and the infliction of pain on others of the human race.

So what does this heritage tell us a sound life philosophy should include? To begin with, one needs self-respect, and to achieve that, one must exercise self-discipline. Moreover, the individual must acquire self-knowledge and the skill to pursue a work in which one can take pride. Furthermore, the individual would want to do the best work possible hopefully to create a thing of utility, beauty, strength, and perhaps of service to others.

Put another way, our sympathetic conscience urges us to pursue a work worthy of our feelings and intelligence. That means being "true to oneself."

In addition, our natural conscience urges us to aid our own kind. Health, security, peace, faith in life for the individual should mean that the necessities, benefits and blessings of life should be apportioned fairly to all.

Each person who has such a vision should share it with those who do not have it. For it is through such faith that we regain our trust in humanity and in ourselves. Our common responsibility is to help all races to the point there no longer is need. This should be our religion in life because we are one humanity.

Stages of Life and Culture

There seem to be stages of creativity and conscience in the individual's life, in "periods" of history, and in the history of culture and civilization.

It may be that our childhood, youth, and vigorous adulthood are primarily creative, although influenced by conscience. The stages are primarily morphogenetic in their sexuality, spirit of adventure, inventiveness, curiosity, and originality, although common sense requires the exercise of symbiotic conscience to prevent us from indulging in excesses that could harm and destroy us.

On the other hand, mature adulthood and advancing age bring with them a greater need for emphasis on symbiotic conscience. It is symbiotic by the yearning to fit things together, to perfect and complete whatever is near and dear to our intelligence or spiritual preoccupations. Mature adulthood is conscience by a greater need to see justice done and to comply with religious commandments. It is conscience by seeing the need for wisdom based on what is natural and moral.

So it may be with periods of cultural history. If at times morphogenetic creativity spills over into immoral excesses in one stage of the Zeitgeist, at another time there may be a symbiotic reaction seeking out temperance, balance, harmony, justice for all, and greater spiritual morality.

Of course, there are phases that weave their way through the artificial compartmentation of time known as "historical periods." These phases act as theses and antitheses. Sometimes they lead to fortuitous syntheses, but sometimes under adverse conditions they decay into decadence, degeneration, and destruction of a society or culture.

In another way of describing such a sequence, there may be fertile, maternal stages of society, which are dominated by the need for propagation. This would be a morphogenetic age, characterized by all manner of ingenuity and innovation as expressed by man and woman.

Such a stage may be superceded by a symbiotic period, where law, rule, regulation, and orthodoxy guide conscience. The time may be consciously patriarchal. Yet, subconsciously the sternness of the period may be mitigated by maternal compassion for the poor and downtrodden. Of course, both intelligent men and women hearken to both sides of human nature, the morphogenetic and symbiotic, the creative and conscientious.

It may even be possible to retrace strands of these two protoprocesses for their cyclical predominance. At times they intertwine; at times they clash. At rare times they may unite to create nobler models of humanity and more evolved stages of human society.

Perhaps some psychologist may enjoy toying with the possibility that the individual moves through life in tandem with morphogenetic passions or symbiotic practicality and wisdom. Perhaps some sociologist may find it worthwhile to research traditional and advanced societies for their morphogenetic and symbiotic stages. Finally, some cultural historian might examine one or a number of related cultures for their parallel morphogenetic and symbiotic periods of creativity and conscience.

The Characteristics of the Creative Conscience

The creative conscience is primarily the consequence of the universal morphogenesis and symbiosis pervasive in nature and in human nature. Each process enables us to grow, add to our knowledge, and optimize one's likelihood of finding and of keeping lifetime happiness.

Morphogenesis: Its Traits and Functions

Morphogenesis:

1. Keeps us in contact with the real world. We need to understand that the Third Millennium has unveiled a new reality.

2. Urges us to explore the world for new knowledge, opportunities, adventures, and discoveries.

3. Encourages us to pursue a healthy curiosity about life and its purpose.

4. Welcomes difficulties, change and problems as a means of testing our emotional maturity and intelligence. Instinctively, we experience accelerated growth and sophistication of mind every time we solve problems and resolve challenges.

5. Makes us stronger, more self-confident, more sure we can master the unexpected in life.

6. Has enormous transformational power. It can inspire our psyche to undergo conversion from a state of self-doubt, helplessness or hopelessness to the reverse. It can energize a new sureness, certainty, and readiness to meet gladly whatever tests our mettle and to help others in need.

7. Can convert the individual from a selfish, self-pitying disposition to a selfless, responsible, and self-reliant character.

Symbiosis: Its Traits and Functions

Symbiosis:

1. Teaches us how to live in the real world. However, we not only adapt to it; we also adapt nature and the world to our own survival needs.

2. Teaches us it takes time and patience for anything vital and viable to take root, grow, flower and bear fruit.

3. Teaches us that self-discipline is the pathway to lifetime survival.

4. Inspires us to study and emulate the perfected designs found in all forms of nature. By such emulation, we teach ourselves the meaning of perfection in our work and creativity.

5. Invites us to mature our plans for life. We need to make decisions, pursue activities, and take actions that will slowly but surely realize the essential desire of our lives.

6. Reminds us to harmonize human relations and to help others live in peace with each other, despite the political, ethnic, cultural or religious differences.

7. Urges us to succor humanity, especially feeding, healing and educating the children of the world.

8. Teaches us we have the inborn creative conscience to shape our personal lives into a meaningful destiny.

AFTERWORD: THE MUTUAL PURPOSE OF NATURE'S PROCESSES AND SOME FINAL CONCLUSIONS

We have presented evidence that a creative conscience exists in nature and humanity.

At times when we catch sight of signs of morphogenesis and symbiosis in life, it makes us wonder if we are witnessing some kind of miracle. If so, what is a miracle?

Usually we think of the term as defining an extraordinary event which evinces divine intervention, or the fulfillment of some spiritual law, or simply an extremely unusual or outstanding event. To be sure, scientists would object that a miracle would imply the suspension of natural law. To our mind, a miracle would somehow surpass what we once designated as natural law. Such a miracle would motivate the human mind to perceive a law higher than what we had hitherto considered possible as a part of reality.

Over years of studying literature, I came more and more to respect the intricate intelligence of great literary works. I came to realize such works did not necessarily spring from the time during which they were written. Imagery, metaphor, irony and fresh insights into human nature evidently arise from a source deeper than our increasingly sophisticated understanding of humanity and its cultures. To be sure, twentieth century research into mythology, depth psychology, anthropology, comparative religion, and sociology has brought us many new and worthwhile perspectives.

Similarly, the study of literature in the context of hermeneutics, rhetoric, criticism, history, art, music, and philosophy did not quite account for the originality and conscience evident in the most unforgettable literary works. As valuable as such cross-references were and as much enlightenment as each field brought with it, something timeless, universal, humane and soul-uplifting was manifest in true masterpieces, but what would that mysterious essence be?

I began to sense that great literature was illumined by the manifestation of some intelligence beyond that of any individual or culture, past or present. That intelligence was not solely demonstrated by geniuses. Somehow it seemed to emanate from Nature itself, but focused by the evolved human mind. Could it be the "eternal soul"? After all, according to tradition, the "soul" permeated the web of life itself. After all, the "soul" was believed to breathe in the spirit of the universe. It motivated humankind to seek out our significance in the cosmos.

But no, not even the human soul accounts for great literature. There is something larger than that, something not only in contact with the infinite but omnipresent in this world.

While searching for a down-to-earth answer, a man unconsciously stood his twelve-inch, wood ruler vertically on its end on his desk. It did not fall. Bemused, he wondered how that was possible. The planet was whirling through space and time and the ruler continued to stand before him as if it were an icon.

Next, he realized he too stood there perfectly balanced among all the dynamic forces of our universe. He too was held between centripetal and centrifugal powers neither crushed to the center of the earth nor flung out into endless space. At this pinpoint of time, he was witnessing a miracle of his existence and of existence itself.

When we pause to reflect on it, all Existence is in equipoise. On earth, we instinctively know that our lives are cradled in Nature and that nature has "always" been there. We know nature has given us our heritage, soundness of body and intelligence. Through a slow but sure evolution over millions of years, nature has educated us to become human beings.

As remarkable as was the realization that his wood ruler stood steadily balanced between powerful, opposing centripetal and centrifugal energies, living nature itself maintained an equilibrium quite as "miraculous." At first, he thought it to be the struggle between life and death, which began in nature at the start of life itself. But no, it was not the struggle for survival. *The miracle was that life had endured.*

Though deeply concealed, equipoise exists at the heart of life itself. In fact, it manifests itself by its own generative power. That power is the effect of two ubiquitous processes with one mutual purpose, which account for all forms of life on earth.

One is morphogenesis, the vital principle in nature that discloses the origin, innovation, development, engendering and birth of all life forms in Nature. (The ancient Greeks were fond of the mythical god Prometheus, who stole the fire of creativity from heaven and gave it to man. On the other hand, they were awed by the mythical god Proteus who could assume any form of life he chose.) The process of morphogenesis finds its presence in mankind by our propensity to adventure, exploration, experimentation, discovery, invention—in sum, humanity's curiosity and creativity.

However, a second force modifies, controls, directs, and guides the indefatigable drive of morphogenesis.

In nature, symbiosis appears to have the responsibility: to coordinate and subordinate all animate activity; to coalesce, concentrate, and combine all vital energies; to organize, methodize, and systematize all life functions; to prioritize, plan and design all distinct life forms; to fuse and integrate the dynamics of every living thing. In short, symbiosis is the creative conscience inherent in Nature.

Mother Nature's wisdom epitomizes it. Our crafts, arts, inventions, literatures, medicine, religions and investigative sciences are morphogenetically generated from her creative nature. Our societies, our synthesizing sciences and philosophies, our established cultures and civilizations emanated from her symbiotic nature. Ageless morphogenesis and symbiosis brought about all human knowledge and what wisdom we have.

To sum up, the dialogue or dialectic between morphogenesis and symbiosis maintains life in the face of natural, accidental and predator death. As such, these universal processes make possible the miracle of life on earth.

Because humankind incarnates Nature's ageless morphogenesis and symbiosis, our timeless intuition tells us that Nature's Creative Conscience is the ultimate source of life that guides the destiny of humanity. In every way that we commit ourselves to this eternal verity, there we emulate the miracle of life itself.

1. The human use of morphogenesis and symbiosis accelerated the evolution of humankind.

2. The co-evolution of morphogenesis and symbiosis is the reason we emerged as Homo sapiens.

3. By the interplay of animus and anima in the individual, each person evolves a more responsible and compassionate being.

4. By the cooperation of man and woman, we mutually educate each other to a fuller understanding of our moral and humane nature.

5. By our educating the young to the wisdom of nature's creative conscience, humankind can evolve beyond its barbarous past. Through time we will engender a world of peace, reciprocity, and mutual respect. We will reach the furthermost, moral evolution possible for humanity.

NOTES

Chapter 1

1. Page numbers in parentheses refer to the Penguin edition of Charles Darwin's *The Origin of Species by Means of Natural Selection or the Preservation of Favoured Races in the Struggle for Life*. Ed. J. W. Burrow. (Harmonsworth, Middlesex, England:1985).
2. Wesson's reservations raise a difficult question: how to achieve a balanced, verifiable understanding of nature's complexities in simple, clear terms that human reason can grasp. If we attempt explanations, established historically on the symmetric logic of philosophy and the physical sciences (viz. induction, deduction, causality, mechanism), our rational language will be ill-equipped to explain the phenomena of life. For this reason the science of biology has had to invent an authentic biologic which aptly and accurately describes and explains natural processes. For this reason human cognition needs to learn to transcend symmetrical logic by understanding philosophically the significance of morphogenetic and symbiotic intelligence to human life.
3. Chapters three through twelve undertake to prove that morphogenesis and symbiosis are responsible for the evolution of human intelligence.

Chapter 2

1. All page numbers in parentheses are from Arthur O. Lovejoy's, *The Great Chain of Being: A Study of the History of an Idea.* (Cambridge, Massachusetts: Harvard UP, 1936/1961).
2. Robinet seems to have anticipated Schelling's *Naturphilosophie* and Bergson's *élan vital* (p. 281).
3. Quotes are from my book, *A Philosophy of Literary Criticism (A Method of Literary Analysis and Interpretation)*. (Jericho N.Y.: An Exposition University Book, 1974).
4. See my *How Nature Taught Man to Know, Imagine and Reason*. (N.Y.: Peter Lang Publishing, Inc. 1995). See 6, "Understanding Metamorphosis," especially pp 101–5 on symbol.
5. Northrop Frye, *Anatomy of Criticism*. "Third Essay, Archetypal Criticism: Theory of Myths." (Princeton, N.J.: Princeton UP, 1957), p. 136.
6. The German theologian-philosopher F.E.D. Schleiermacher held a conception of humanity close to that of the English romantic poets. His basic premise was that "the fundamental characteristic of nature aims at diversity (*Mannigfältigkeit*) and individuality, (*Eigentümlichkeit*) (Lovejoy, p. 308). Moreover, as did the romantics in England, he viewed morality "as all encompassing sympathy." But beyond one-on-one empathy, he meant that the truly conscientious individual should seriously seek to learn "the character of various

cultures and mankind in all periods of history." (pp. 308–9) Moreover, no one should neglect to cultivate his or her own uniqueness. (p. 310) In sum, Schleiermacher's outlook embraced all humanity, for he believed in "the intrinsic worth of human diversity." (p. 313)

Chapter 3

1. G.I. Edwards. *Biology, the Easy Way*. (Hauppauge, N.Y.: Barron's Educational Series, second edition, 1990).
2. Ibid.
3. Darwin's *Origin of the Species*. (Harmsworth, Middlesex, England, 1985). It may be of interest to compare the stages of differentiation and integration with C.G. Jung's description of individuation through self-knowledge.

Chapter 4

1. Victor B. Scheffer, *Spires of Form, Glimpses of Evolution*. (N.Y. : Harcourt, Brace & Jovanovitch, 1985). All page numbers in parentheses are from Scheffer's book.
2. Gabrielle I. Edwards, *Biology, The Easy Way*. (Hauppauge, N.Y.: Barron's Education Series, second edition, 1990). (p. 21)
3. Matt Ridley, *Genome. The Autobiography of a Species in 23 Chapters*. (N.Y.: Harper Collins, 2000).
4. L. Thomas, *The Lives of a Cell. Notes from a Biology Watcher*. (N.Y.:Bantam Books, 1975). All page numbers in this section are from Thomas's book.
5. S.A. Wainwright, *Axis and Circumference. The Cylindrical Shape of Plants and Animals.* (Cambridge, MA.: Harvard UP, 1988). All page numbers in this section are from Wainwright's book.
6. The examples of morphogenesis and symbiosis in external and microscopic nature are taken from diverse published sources.
7. R. Wesson, *Beyond Natural Selection*, (Cambridge, MA.: MIT P,1991). All pages in parentheses in this section refer to this book.
8. *Merriam Webster's Collegiate Dictionary*. Tenth Edition. (Springfield, MA. 2001).

Chapter 6

1. The history of ideas illustrates these polarities running through humanity's intellectual history. That history corroborates in no uncertain way the fact that our own evolution has been due to a dialectic of extreme polar viewpoints.

Chapter 7

1. R. J. Richards, *The Meaning of Evolution: The Morphological Construction and Ideological Reconstruction of Darwin's Theory*. (Chicago & London: U of Chicago P, 1992). Page numbers in parentheses in this section refer to this book.
2. Barnhart & Steinmetz, *The American Heritage Dictionary of Science*. (Boston, MA. : Houghton Mifflin, 1986). Page numbers in parentheses in this section refer to this book.
3. When Oedipus was asked which animal went on three legs, the ancient Greek replied, "Man. As a baby he crawls on all fours and when an adult he walks on two legs. But as an old man, he needs a staff to get through the rest of his life. That is his third leg."

Chapter 8

1. All time references to past civilizations, events and examples of human nature are based on Dr. Frank P. King's *A Chronicle of World History, From 130,000 Years Ago to the Eve of A.D. 2000*. (Lanham, Maryland: UP of America, 2002).

Chapter 9

1. The time approximations come from Frank P. King's, *A Chronicle of World History: From 130,000 Years Ago to the Eve of AD 2000*. (Lanham/New York/Oxford: UP of America, 2002.
2. All page numbers in parentheses refer to Frank King's, *A Chronicle of World History*.
3. For a description of the signs of the zodiac, see *MerriamWebster's Collegiate Dictionary. Tenth Edition*. (Merriam Webster's, Inc. Springfield, MA. , 2001). p. 1372.
4. For a description of Gregorian, Jewish and Islamic calendars, see the same *Collegiate Dictionary*, p. 753.

Chapter 10

1. All page numbers in parentheses refer to W. Lillie's *Introduction to Ethics*. London, Methuen, 1961.
2. See Eric Chaisson. *Universe. An Evolutionary Approach to Astronomy*. (Englewood Cliffs, N. J., Prentice Hall, 1988.On "Climate," pp. 505–06.

Chapter 11

1. This summary statement is a quotation from my book, *Beyond Literary Theory: Literature as a Search for the Meaning of Human Destiny*. (Lanham, Maryland: UP of America, 2001), p. 158.
2. All translations from Diel's *Psychologie de la Motivation* are my own.
3. Compare the parasympathetic and sympathetic nervous systems as correlated to metaphor and irony and to Hegel and Socrates' use of dialectic.

Chapter 12

1. Primitive species such as bacteria and bluegreen algae do not have an organized nucleus (cells known as *prokaryotes*.), but most cells do have such a nucleus, classified as *eukaryotes*. Its nucleolus is the site of the synthesis and storage of RNA (ribonucleic acid). (Barron's, second, edition, p. 35)

GLOSSARY

The following glossary is limited to the major ideas which this book undertakes to convey. Many other biological terms are public knowledge and do not need any further definition than what appears in a college dictionary.

Key Philosophical Concepts Prior to Darwin's Theory

Being

A modern understanding of being defines the concept as describing the totality of things in existence or in life itself. A second definition describes the qualities or essence that a being embodies.

Historically, philosophers and theologians have undertaken to describe the being of nature as ordered, composed and organized according to God's will. In fact, ever since the ancient Greeks, reason was considered the controlling principle of the universe. Thus humankind's highest attainment was to be rational. (For further discussion, see the *principle of plenitude* in Chapter 2.

Becoming

Today the meaning of becoming is that life forms come into existence, change and develop. However, in ancient Greece, Aristotle's *De Anima* (fourth century B.C.E.) already questioned the logical classification of forms in animate nature and doubted that nature could be explained in any rational order. In later centuries there arose a clearer understanding of nature's perpetual becoming. Increasingly, thinkers noted the variations and purposeful adaptations of plants and animals. (For further explanation, see the *principle of continuation* in Chapter 2.)

The German philosopher Leibniz (1646–1716) interpreted natural forms as having emanated from God, that is as monads possessing degrees of perfection. To following philosophers, becoming seemed to account for the evolution of evolution. (For additional examples, see the *principle of gradation* in Chapter 2.)

Key Terms Used in Darwin's Theory

On the other hand, Darwin's theory of evolution (1859) presented a solid scientific argument that nature's various types of plants and animals had their origin in preexisting archetypes and that the generation of species explained their distinguishable characteristics, variations and mutations. Among his concepts are the following.

The Struggle for Existence

Species compete for food, space and mates. Those individuals less capable of survival, with limited capacities or ineffective strategies, tend to be eliminated by their competitors. Moreover, those that do survive pass on their successful traits to offspring and descendants.

Natural Selection

Those individuals or groups in nature best adapted to their environment have the greatest chance of survival and of reproducing their own kind. This natural process allows them to perpetuate the genetic characteristics best suited to that environment.

Survival of the Fittest

The expression is synonymous with natural selection.

Variation

Divergence in the characteristics of an organism, from those typical of its species, led to variation. The variety of species in the world is partly a consequence of their adaptation to various climates and environments.

On the other hand, Darwin thought species underwent a process of convergence as they evolved similar physical traits, structures and habits due to their occupying the same environment.

Key Terms Used in Twentieth Century Ecology and Microbiology

Ecology

A branch of biological science, ecology studies the relationship between organisms and their environment. Its aim is to be able to achieve a holistic description of their relations and patterns of existence.

Microbiology

Microbiology is a relatively recent branch of biology which studies microscopic forms of life as cells and genes. Once unseen by the naked eye, the activity of single cells is now understood to be responsible for integrating all multicellular organisms through a process of ever increasing complexity of organization. Indeed, the process of consolidation effects the growth of tissue, muscle and bone as well as the regulation of all intrasomatic processes that characterize a life form.

Genetics

A new branch of microbiology is genetics or the study of the gene. As a sequence of nucleotides in DNA and RNA located in the germ plasm, the gene transmits our traits and characteristics from generation to generation. Obviously this new science extends our understanding of all life's intrinsic natural processes.

Key Terms Used in The Creative Conscience (Third Millennium Biology)

Morphogenesis

Scientifically speaking, morphogenesis is the formation and differentiation of tissues and organs. In other words, the process is responsible for the origin and development of bodily organs, tissue and bone.

However, as discussed herein, morphogenesis is seen as encompassing much larger aspects and processes of nature, especially as nature has directly influenced the evolution of humankind. Morphogenesis not only activates and actualizes the embryonic development of the individual human being but also directs the ontogenesis of all human life through its stages of infancy, childhood, adolescence, and adulthood into advanced age. By extension, the morphogenetic nature of our species is shown to account for humankind's creativity. Further, morphogenesis has guided humankind to create its worldwide cultures and civilizations.

Symbiosis

The usually accepted definition of symbiosis emphasizes the ecological relationship between organisms. Figuratively speaking, symbiosis describes the tolerant association and living together of two dissimilar organisms. In other words, they are mutually dependant on each other as in mutualism.

A microbiological interpretation makes clear that symbiosis is manifest in the arrangement, organization and integration of cells into multicellular tissues, organs, and entire life support systems.

However, as morphogenesis itself, symbiosis plays a vital role not only in the development of the individual human being but also in human evolution. Closer examination of symbiosis reveals it is responsible for the design and form evident throughout animate nature. Moreover, it explains the emergence in humankind of a nature-born conscience. Indeed, it may be called a symbiotic conscience, which propels us to complete and perfect whatever we undertake to do. It also explains our capacity of compassion for the distressed, the oppressed, the sick, the needy, and the helpless.

In sum, morphogenesis and symbiosis constitute the basis for humankind's biogenetic intelligence. Over time these natural processes have interacted dialectically to evolve the creative conscience of the human race.

BIBLIOGRAPHY

On General Nature

Attenborough, David. *The Atlas of the Living World*. Boston: Houghton, 1989.
Bailey, Kenneth. Ed. *Encyclopedia of Nature*. N.Y.: Gallery, 1973.
Burton, Robert. *The Mating Game*. London: Peerage, 1976.
Caufield, Catherine. *In the Rainforest. Report from a Strange Imperiled World*. Chicago: U of Chicago P 1984.
Edinger, J. G. *Watching for the Wind. The Seen and Unseen Influences on Local Weather*. N.Y.: Doubleday, Anchor Books, 1997.
Fifield, Richard. Ed., *The Making of the Earth*. N.Y.: Basil Blackwell, 1985.
Halpern, Daniel. Ed. *On Nature, Landscape, and Natural History*. San Francisco: North Point, 1987.
Levenson, Thomas. *Ice Time, Climate, Science and Life on Earth*. N.Y.: Harper, 1989.
Owen, Denis. *Camouflage and Mimicry*. Chicago: U of Chicago P, 1982.
Platt, Rutherford. *This Green World*. N.Y: Dodd, 1988.
van der Wall, S.B. *Food Hoarding in Animals*. Chicago: U of Chicago P., 1990.

Science, Evolution, Ecology, Microbiology

Augros & Stanciu, *The New Biology. Discovering the Wisdom of Nature*. Boston and London: Shambala, 1988.
Barnhart & Steinmetz. *The American Heritage Dictionary of Science* Boston, Mass.: Houghton Mifflin, 1986
Boucher, D.H. Ed. *The Biology of Mutualism, Ecology and Evolution*. N.Y.: Oxford UP, 1988
Darwin, Ch. *The Origin of Species by Means of Natural Selection*. Chicago, London and Toronto: Encyclopedia Britannica, 1952.
_____. *The Descent of Man*. Chicago, London and Toronto: Encyclopedia Britannica, 1952.
_____. *The Origin of Species by Means of Natural Selection or the Preservation of Favoured Races in the Struggle for Life*. Ed. J. W. Burrow. Harmondsworth, Middlesex, England, 1985.
Goldsmith, E. *The Way. An Ecological World View*. Boston, Mass. Shambala, 1993.
Greenstein, G. *The Symbiotic Universe. Life and Mind in the Cosmos*. N.Y.: William Morrow, 1988.
Horgan, J. *The End of Science. Facing the Limits of Knowledge in the Twilight of the Scientific Age*. N.Y.: Broadway Books, 1997.
Lazlo, E. *Evolution. The Grand Synthesis*. Boston: Shambala, 1987.

Margulis & Fester. Eds. *Symbiosis as a Source of Evolutionary Innovation: Speciation and Morphogenesis.* Cambridge, MA. MIT P, 1991.
Nitecki, M. H. Ed. *Evolutionary Innovations.* Chicago & London: U of Chicago P, 1990.
Richards, R.J. *The Meaning of Evolution. The Morphological Reconstruction and Ideological Reconstruction of Darwin's Theory.* Chicago & London: U of Chicago P, 1992.
Singer, Ch. *A History of Scientific Ideas. From the Dawn of Man to the Twentieth Century.* N.Y. Dorset P., 1959.
Scheffer, V.B. *Spires of Form. Glimpses of Evolution.* N.Y.: Harcourt, Brace, Jovanovich, 1985.
Smith, J.M. *The Theory of Evolution.* N.Y. Penguin, 1979.
Storer, J.H. *The Web of Life (A First Book of Ecology).* N.Y.: New American Library, 1956.
Thomas, L. *The Lives of the Cell. Notes of a Biology Watcher.* N.Y.: Bantam Books, 1975.
Thomson, K.S. *Morphogenesis and Evolution.* N.Y. & Oxford: Oxford UP, 1988.
Wainwright, S.A. *Axis and Circumference. The Cylindrical Shape of Plants and Animals.* Cambridge, Mass. Harvard UP, 1988.
Wesson, R. *Beyond Natural Selection.* Cambridge, Mass.: A Bradford Book, MIT Press, 1991.

The Human Body and Mind

Achterberg, Jeanne. *Imagery in Healing. Shamanism and Modern Medicine.* Boston: Shambala, 1989.
Ackerman, Diane. *A Natural History of the Senses.* N.Y. Random, 1990.
Assimov, Isaac *The Human Brain. Its Capacities and Functions.* N.Y. New American Library, 1965.
Bateson, Gregory. *Mind and Nature. A Necessary Unity.* N.Y.: Bantam, 1988.
Baumgart, Ernest. *La Vision.* Paris, France: P U de France, 1968.
Bouthoul, Gaston. *Les Mentalités.* Paris, France: P U de France, 1971.
Brunschvicg, Leon. *Les Ages de l'Intelligence.* Paris, France: P U de France, 1953.
Bucke, R.M. *Cosmic Consciousness. A Study of the Evolution of the Human Mind.* Secaucus, N.J.: Citadel, 1977.
Donald, Merlin. *Origins of the Modern Mind. Three Stages in the Evolution of Culture and Cognition.* Cambridge, Mass.: Harvard UP, 1991.
Ferguson, Marilyn. *The Brain Revolution. The Frontiers of Mind Research.* N.Y.: Bantam, 1975.
Gardner, Howard. *The Quest for Mind. Piaget, Levi-Strauss, and the Structuralist Movement.* N.Y.: Random, 1974.
Gazzaniga, M.S. *Mind Matters. How Mind and Brain Interact to Create Our Conscious Lives.* Boston: Houghton, 1988.
Ghandi, Kishore. *The Evolution of Consciousness.* N.Y.: Paragon, 1986.
Langer, Ellen J. *Mindfulness,* N.Y. : Addison-Wesley, 1989.
Lewis, Paul & Rubenstein, David. *The Human Body.* N.Y.: Bantam, 1983.
Litvak, S. and Senzee, A.W., *Toward a New Brain. Evolution and the Human Mind.* Englewood, N.J.: Prentice Hall, 1986.
Lloyd, Dan. *Simple Minds.* Cambridge, Mass,: MIT P, 1989.
Piaget, J. *Biologie et Connaissance.* Paris, France: Gallimard, 1967.
Pickering, J. & Skinner, M. Eds. *From Sentience to Symbols. Readings on Consciousness.* Toronto & Buffalo: U of Toronto P, 1990.
Pool, J. Lawrence, M.D. *Nature's Masterpiece. The Brain and How it Works.* N.Y. Walker, 1986.
Strauch, E.H. *How Nature Taught Man to Know, Imagine and Reason. (How Language and Literature Recreate Nature's Lesson.)* N.Y.: Peter Lang Publishers, 1995.

Thinking, Creativity and Conscience

Bernes, Jeanne. *L'Imagination*. Paris, France: P U de France, 1975.
Boirel, René. *L'Invention*. Paris, France: P U de France, 1972.
Bolton, Neil. *The Pyschology of Thinking*. London: Methuen, 1972.
Bridoux, Andre. *Le Souvenir*. Paris, France: P U de France, 1970.
Carpentier, Raymond. *La Connaissance d'Autrui*. Paris, France: P U de France, 1968.
D'Arcy, Philippe. *La Reflexion*. Paris, France: P U de France, 1972.
Davy, M. M. *La Connaissance de Soi*. Paris, France: P U de France, 1972.
De Bono, Edward. *New Think. The Use of Lateral Thinking in the Generation of New Ideas*. N.Y.: Avon, 1971.
Dimnet, Ernest. *The Art of Thinking*. Greenwich, CT.: Fawcett, 1956.
Granger, Gilles Gaston. *La Raison*. Paris, France: P U de France, 1974.
Hawkes, Terence. *Metaphor*. London: Methuen, 1977.
Jackson, K.F. *The Art of Solving Problems*. London: Heinemann, 1975.
Jung, C. G. . *The Archetypes and the Collective Unconscious*. Trans R. F.C. Hull, Princeton, NJ: Princeton UP, 1959.
Khatena , Joe. *Imagery and. Creative Imagination*. N.Y.: Bearly, 1984.
Koestler, Arthur. *The Art of Creation*. N.Y.: Dell, 1967.
Luthi, Max. *Märchen*. Stuttgart, Germany: Metzler, 1964.
Muecke, D.C. *Irony*. London: Methuen, 1970.
Parnes, Sidney J. *Creativity: Unlocking the Human Potential*. Buffalo, N.Y.: D.O.K., 1972.
Rouquette, Michel Louis. *La Créativité*. Paris, France: P U de France,1973.
Ruthven, K.K. *Myth*. London: Methuen, 1976.
Strauch, E.H. *Creative Writing for Africans*. Lanham, Maryland: UP of America, 1995.
Veraldi, Gabriel and Brigitte. *Psychologie de la Création*. Paris France: Centre d'Étude, 1972.

Humankind and Religion

Armstrong, K. *A History of God. The 4,000 Year Quest of Judaism, Christianity, and Islam*. N.Y. Ballantine Books, 1994.
Campbell, I.C. *A History of the Pacific Islands*. Berkeley: U of California P, 1989.
Diel, Paul. *le symbolisme dans la mythologie grecque*. Paris, France: Petite Biblothèque Payot, 1966.
_____. *La Divinité. Le Symbole et sa signification*. Paris France: Petite Bibliothèque. Payot, 1971.
_____. *le symbolisme dans la bible. l'universalité de langage symbolique et sa signification psychologique*. Paris, France: Petite Bibliothèque Payot, 1975.
Eliade, Mircea. *A History of Religious Ideas. From Stone Age to the Eleusinian Mysteries*. Trans. Willard F. Trask. Chicago: U of Chicago P, 1978.
Fagan, Brian. *Kingdoms Of Gold, Kingdoms of Jade. The Americas Before Columbus*. N.Y.: Thames & Hudson, 1991.
Flornoy, Bertrand. *The World of the Inca*. Trans. Winifred Bradford. Garden City, N.Y.: Doubleday, 1958.
Forde, Daryll. *African Worlds. Studies in the Cosmological Ideas and Social Values of African Peoples*. Great Britain: Oxford UP, 1991.
Howells, W. *Back of History. The Story of Our Origins*. Garden Clty, N.Y.: Anchor Books, 1963.
King, Frank P. *A Chronicle of World History from 130,000 years ago to the eve of A.D. 2000*. Lanham, New York, Oxford: UP of America, 2002.

Peuch, Henri Ch. Ed. *Histoire des Religions. La Formation des Religions Universelles et les Religions de Salut.* Paris: Encyclopédie de la Pléiade, Gallimard, 1970.
Prescott, William. *The World of the Aztecs.* Genève Paris: Éditions Minerva, 1990.
Reichard, Gladys A. *Navaho Religion. A Study of Symbolism.* Princeton, N.J.: Princeton UP, 1977.
Sailbull, olle Solomon and Carr, Rachel. *Herd and Spear. The Masai of East Africa.* London: Collins & Harvill, 1981.
Schele, Linda and Freidal, David. A *Forest of Kings. The Untold Story of the Ancient Maya.* N.Y.: William Orrow, 1990.
Sumner, W.G. *Folkways. A Study of the Sociological Importance of Usages, Manners, Customs, Mores, and Morals.* N.Y.: Ginn, 1940.
Thomson, O. *A History of Sin.* 'N.Y.: Barnes and Noble, 1995.
Yenne, Bill. *Encyclopedia of North American Indians. A Comprehensive Study of Tribes from the Abitibi to the Zuni.* Greenwich, CT. Crown, 1986.

Philosophy

Boschenski, J.M. *Introducción al Pensamiento Filosófico.* Barcelona, Spain: Herder, 1986.
Dilthey, Wilhelm. *Théorie des Conceptions du Monde. Essai d'une Philosophie de la Philosophie.* Trans. Louis Souzin, Paris, France: P U de France, 1946.
Foster, David. *The Philosophical Scientists.* N.Y.: Dorset, 1991.
Labax, Emile. *La Dialectique du rhythme de l'Univers.* Paris France: Librairie Philosophique J. Vrin, 1925.
Lillie, W. *Introduction to Ethics.* London: Methuen, 1961.
Lovejoy, A.O. *The Great Chain of Being.* Cambridge, Mass.: Harvard UP, 1936/1961.
Magill, Frank N. Ed. *Masterpieces of World Philosophy in Summary Form.* N.Y.: Harper, 1961.
Makreel, Rudolf A. *Dilthey, Philosopher of the Human Studies.* Princeton, N.J.: Princeton UP, 1975.
Morot, Sit. Ed. *La Pensée Negative.* Paris, France: Aubier, 1947.
Reymond, A. Virieux. *Introduction à l'Épistemologie.* Paris, France: P U de France, 1972.
Rothacker, Erich. *Logic und Systematik der Geisteswissenschaften.* Bonn, Germany: Bouvier, 1957.
Strauch, E.H. *A Philosophy of Literary Criticism. (A Method of Analyzing and Interpreting the Human Experience in Literature.)* Jericho,N.Y.: Exposition Press UP, 1974.
_____. *brief eternity* (an existentialist novel). Huntington, Virginia. University Editions, 1995.
_____. *Beyond Literary Theory. Literature as the Search for the Meaning of Human Destiny.* Lanham, Maryland: UP of America, 2001.
Wiener, Philip P. Ed. *Dictionary of History of Ideas. Psychological Ideas in Antiquity to Zeitgeist.* N.Y. Scribner, 1973.

INDEX

To the Reader

The *Creative Conscience as Human Destiny* presents a third millennium philosophy of evolution. It endeavors to explain nature's universal processes as they manifest themselves in human nature. A philosophical exposition calls for a topical alpha index to help the reader more readily grasp the central ideas.

Based on a post-Darwinian theory of nature, of human evolution and the humanization of our species, the book undertakes to demonstrate how nature's Creative Conscience finds expression in humankind's own ingenuity, knowledge and cultures.

A topic index enables the interested reader to review holistically any area of special interest. Hence the reader will note that each broad concept provides a list of generally well known names and concrete examples to illustrate each basic idea. All main concepts are themselves alphabetized, and within each of the twelve sections of the index, the sub-ideas are also alphabetized.

Thus, the purpose of the topical index is twofold: (1) to facilitate reader understanding of the global significance of the creative conscience both in Nature and in humankind, and (2) to survey more effectively those areas of knowledge the reader may wish to investigate in breadth and depth.

༄༅༔

Topical Alpha Index

I	Being and Becoming
II	Darwin's Theory of Evolution
III	Ecology and Microbiology
IV	Human Ingenuity and Culture
V	Human Intelligence
VI	Humanization of Our Species
VII	Irony as Fate, Metaphor as Destiny
VIII	Nature's Forms of Intelligence
IX	Nature's Polar Processes
X	Polarity of Human Nature in Culture
XI	Professors, Sages, and Scientists
XII	Religions and Life Philosophies

I. Being and Becoming

becoming xvi, 19, 22, 25-26, 33-34, 53
being xvi, 19, 22, 25-26, 33-34, 53
philosophy prior to Darwin
    Leibniz 24-26
    Lovejoy's history of ideas
        19, 21, 24-26, 34
    metaphor as metaphysical meaning
        28-30
    Schiller, F. 30-32
principle of continuity, 21
    Aristotle 21
    Kant, I. 21
    Robinet, J.B. 21

principle of gradation 23
    Aristotle 22
    Dante A. 22
    Neoplatonists 22
principle of plenitude 19-21
    Aquinas 20
    Augustine 20
    Averroës 20
    Bruno, G. 21
    Neoplatonists 21
    Plato 19-20
    PseudoDionysius 20

II. Darwin's Theory of Evolution

causality 107, 180
convergence xvi, xvii, 39-40, 43
divergence xvi, xvii, 4, 6, 37, 41, 43
extinction 3, 4, 5, 184
evolution 10, 14-15, 107, 109, 120,
    122, 124
mechanisms 107, 179-180, 205
nature's web of complex relations 3, 14

natural selection 4, 15, 120, 230
Origin of Species 3, 226
struggle for survival 4, 41, 107-108,
    184, 230
variation 4-6, 179, 230

III. Ecology and Microbiology

cell 7, 10
co-evolution 6, 52, 55
double helix 55-56
ecology xv, xvi, xix, 3-4, 6, 9-11, 14, 31, 33, 64, 180-181, 183-184, 231
gene (RNA/DNA) 8, 55-56, 66-69, 231
genome intelligence 69
homeostasis 66, 71-73
microbiology xv, xvi, xix, 3-4, 6-8, 10- 11, 14, 30-31, 33, 55, 65, 69, 108, 110, 180-181, 183-184, 231
neoDarwinists 8, 9, 14, 68
Scheffer, J.B. 50-55, 226
Thomas, L. 56-57, 226
Wainwright, S.A. 38, 57-59, 115, 226
Wesson, R. 8, 12-14, 64-68, 226

IV. Human Ingenuity and Culture

alphabets and writing systems 133-4, 144
arches, columns, domes, vaults 145
architecture 136
arithmetic, geometry 136, 144
calendars 133, 144
carved utensils and vessels 140
cave sketches and painting 140
centers of learning 135
centers of worship 136
central market places 136
clay, use of 140
circumnavigation of globe 137
civic liberties and duties 147
construction: roads, bridges, harbors, canals 133
cooking 145
counting and measuring systems 142
craftsmen and artisans 145
domestication of animals 141
drainage systems 136, 146
durable shelters 142
education 138
engineering 149
fishing 142
food cultivation 142, 145
hides into clothing and shelter 141
hunting 139
interpreters of knowledge 135
inventions and innovations 146, 149
irrigation 143
languages xvii, 130, 133, 144
laws and rights 147
libraries 135
cloth from plants 145
megaliths 145
metallurgy 143
migrations 141
navigation 143
number systems 145
potter's wheel 143
pottery 146
record keeping 133
sewers 146
signs of the zodiac 144
smelting 143
staple crops 142
stone walls and towers 142
systems of counting 133
tombstones 140
tools and weapons 141
translation 135
trade 146
weights and measures 135-6 (standardization)
women's domestic creativity 146

V. Human Intelligence

creative (morphogenetic) intelligence
xvii-xviii, 15, 27, 30-31, 33, 70, 83-84, 87-88, 110, 120, 122-125, 181, 183-184, 186-189, 190-193, 195-198, 205-206, 208-210, 213-216, 218-220, 231-232

conscience as (symbiotic) intelligence
xvii-xviii, 15, 27, 30-33, 45, 64, 70, 83, 87-88, 109-110, 121-123, 125-126, 150, 181, 184, 186-197, 199, 205-206, 208-210, 212-214, 217-220, 231-232

cultural intelligence 93-97, 218
  Aeschylus 95, 194
  Boccaccio 135
  Buddha 158, 162
  Dante 19, 22, 135
  Diel, P. 197-200
  Euripides 95, 193
  Gandhi, Mahatma 168
  Goethe, J.J. von 30
  Hegel, G.W.F. 172, 176, 202-203, 227
  Jesus 152, 201
  Jung, C.G. 118, 136
  Lillie, W. 169-171
  Lovejoy, A.O. 19, 22, 26-28, 31, 44, 151
  Moses 152, 157
  Muhammad 156
  Pascal, Blaise 47, 171, 214
  Petrarch 135
  Rousseau, J.J. 168
  Schleiermacher, F.E.D. 225
  Schiller, F. 19, 30-31

  Sophocles 95, 193

philosphical intelligence 16, 103
  Aquinas 19-20
  Aristotle xviii, 19, 21-22, 97, 154, 169, 172
  Augustine 19-20
  Averroës 19-20
  Bergson, H. 171-172
  Bruno, Giordano 19, 21
  Descartes, R. 10, 44, 180
  Epicurus 155
  Green, J.H. 198, 172-3
  Hegel, G.W.F. 172, 176, 202-203, 227
  Kant, I. 21, 44
  Laotzu 151
  Leibniz, G.W. 19, 24-25
  Locke, J. 160
  Lovejoy, A.O. 19, 22, 26-28, 31, 44, 150
  Neoplatonists 19-20
  Pascal, Blaise 47, 171, 214
  Plato xviii, 19-21, 153-154, 169, 176, 203
  Socrates 95, 153-154, 176, 203, 227
  Sophists 203
  Zeno of Citium, Cyprus 155

Problem-solving intelligence xviii, 98-99

Rational intelligence 70-80, 82, 85, 97-99

Survival intelligence 79-83

VI. Humanization of Our Species

Note: Section numbers refer to topical alpha index.

Human Ingenuity and Culture, see section IV
Human Intelligence, see section V
  co-evolution 208-210

conscience as (symbiotic) intelligence, all pages
creative (morphogenetic) intelligence, all pages

male and female, dialectic as evolution 197
  as co-evolution 199
man and woman, archetypal destinies 194, 196
  masculine character and feminine personality 196, 199
  (animus) and (anima)
monotheism xviii, 161
mothers 80-81
Nature's Three Principles, see section I, Being and Becoming
Past Life Philosophies, see section XII
Chapter 10, Third Millennium Philosophy
Professors, Sages and Scientists, see section XI
psychomachia 30, 41, 139, 156
reciprocity 151, 161
Religion and Life Philosophies, see section XII
sōphrosynō, 194, 211
subconscious 197, 201
  (Paul Diel)
Supraconscience 197-200
  (Paul Diel)

VII. Irony as Fate, Metaphor as Destiny

irony 200-203
ironic intelligence 95
  Aeschylus 95, 194
  Euripides 95, 193
  Socrates 95, 153-154, 169, 176, 202-203
  Sophists 203
  Sophocles 95, 193
irony in ancient tragedy
  hamartia (blindness) 95
  hybris (pride, arrogance) 95
sōphronsynē (moderation) 194, 211
metaphor 28-29, 34, 200-202
metaphoric intelligence 95-96
  Dante, A. 19, 22, 135
  Goethe, J.J.von 30
  Hegel, J.W.F. 122, 176, 202-203, 229
  Keats, J. 29
  Lovejoy, A.O. 19, 22, 26-28, 31, 44, 151
  Schiller. 19, 30-31
  Schleiermacher, F.E.D. 225
  Wordsworth, W. 29
metaphor and irony as keys to human destiny 200-203
personification 29
symbol 28-29
  archtypes 28, 118, 136
  Diel, Paul 197-200
  Jung, C.G. 118, 136

VIII. Nature's Forms of Intelligence

biogenetic intelligence 3, 11, 13, 16, 63-65, 69-70, 72-73, 123, 196-197, 205, 223-224
creative conscience xv-xviii, 42, 47, 138, 185, 187, 190, 192-193, 199, 205, 208, 216, 219, 223
evolutionary intelligence 22, 31, 33-34, 41-42, 47, 62-63, 65-72, 83, 86-89, 103-105, 112-113, 121-123, 185, 197, 199-200, 210-211, 223, 226, 232
holistic intelligence xvi, 100-101
homeostatic intelligence xvi, 9, 45, 101
homologous intelligence xvi, 37, 99-100
morphogenetic (creative) intelligence xv-xvi, 3, 7, 11-16, 32-34, 38, 45, 49-62, 64-67, 87-88, 111, 116, 118-121, 124, 181, 183-184, 188-190, 192-193, 198, 200-209, 211, 213-215, 219, 222-223, 231-232

ontogenetic intelligence 65, 69, 103,
  181
symbiotic intelligence (as conscience)
  xv-xvi, 3, 8-9, 11-16, 31-34, 45,
  49, 51-59, 62-65, 79-80, 87-88,
  111, 116, 121, 125, 181, 183,
  185-188, 192-194, 198, 206-207,
  211-215, 223, 232
survival intelligence xvi, 79-83

IX. Nature's Polar Processes

complexification 42-43, 89-90, 114,
  116-117
convergence 110, 114-116, 121-124,
  129, 160, 176
divergence 110-111, 114-115, 121-123,
  160, 176
dualism versus polarity 44-45

polarity and polar processes 45-46, 48
simplification 42-43, 89-90, 109-110,
  114, 117-118, 121-122
mutual purpose of polarities 49, 52, 57,
  179

X. Polarity of Human Nature in Culture

convergence 130
  alphabets 133-134
  architecture 136
    (mosques, cathedrals)
  arts, literature and philosophy 137
  astronomy 134
  calendars 133-134
  central marketplace 136
  centers of learning 135
  centers of religious and cultural
    activities 135
  circumnavigation of globe 137
  construction practices 133
  convergence of aptitudes, talents
    and intelligences 138
  education 138
  ethical convergence 135
  interpreters of literature, philoso-
    phy and sciences 135
  languages 130
  libraries 135
  mathematics 134

pilgrimages 136
public projects 133
record keeping 133
religion 131-132
scholarship 135
standardization of weights and
  measures 135-136
survival practices 133
systems of counting 133
translations 134
writing systems 133-134
divergence, fanaticism,
  fundamentalismism 137
language 130
religion 131-132, 137-138
witchcraft trials 131
war 131-132
mutual purpose 134
  common purpose 136-137
  moral convergence 137
  mutual mortal destiny 138

XI. Professors, Sages and Scientists

Barlow, C. 200
Berdyaev 171
Confucius 151

Copernicus 21
Darwin, Charles xvi, xvii, 3-6, 10, 13-
  15, 37, 39-41, 43, 107-109, 120,

122, 124, 160, 167, 179-180, 226, 230
Diderot, D. 159
Diel, Paul 197-200
Freud, S. 118, 197
Frye, Northrope 29
Galileo 179
Gould, Stephen 12
Green, T.H. 108, 172-173
Jung, C.G. 118, 198
Kepler, J. 148
Leeuvenhoek, A von 148
Linnaeus, C. 42
Lovejoy, A.O. 19, 22, 26-28, 31, 44, 150

Morgan, L. 170
NeoDarwinists 8, 9, 68
Newton, I. 148, 179
Eldridge, Niles 12
Reid, L.A. 171
Richards, R.J. 37, 108-109
Robinet, J.B. 19, 21
Scheffer, V.B. 50-55, 226
Spencer, H. 189
Thomas, L. 56-57, 226
Wainwright, S.A. 38, 57-59, 115, 226
Wesson, R. 64-68, 226
Whittaker, R. 42-43

XII. Religions and Life Philosophies

Brahmanism 158-9, 162
Buddhism xviii, 157-8
Christianity xviii, 30, 155-6, 161 , 201
Confucianism xviii
Deism 160
Enlightenment 159
Epicureanism xviii, 155, 162
Hinduism xviii, 157-8, 161
Islam xviii, 155-7, 161
Judaism xviii, 152, 155-6, 161 201
Manichaeism 161
Mayhayana 158

monotheism xviii, 161
morphogenetic ideal of life xviii, 168-172, 176, 184, 189, 209, 219-220
religion xvii, 150
Stoicism xviii, 156, 160, 162 Sufism 157
Supraconscience 199 197-200
symbiotic ideal of life xviii, 169-171, 176, 179, 184, 189, 220
Taoism xviii, 150-151 Zoroastrianism 155, 161